METHODS IN MOLECULAR BIOLOGY™

Series Editor
John M. Walker
School of Life Sciences
University of Hertfordshire
Hatfield, Hertfordshire, AL10 9AB, UK

For further volumes:
http://www.springer.com/series/7651

miRNA Maturation

Methods and Protocols

Edited by

Christoph Arenz

Institute for Chemistry, Humboldt-Universität zu Berlin, Berlin, Germany

Editor
Christoph Arenz
Institute for Chemistry
Humboldt-Universität zu Berlin
Berlin, Germany

ISSN 1064-3745 ISSN 1940-6029 (electronic)
ISBN 978-1-62703-702-0 ISBN 978-1-62703-703-7 (eBook)
DOI 10.1007/978-1-62703-703-7
Springer New York Heidelberg Dordrecht London

Library of Congress Control Number: 2013949785

Printed on acid-free paper

Humana Press is a brand of Springer
Springer is part of Springer Science+Business Media (www.springer.com)

Preface

The last decade has seen a dramatic development in the field of micro RNAs (miRNAs). Starting with a small set of small noncoding regulatory RNAs in *D. melanogaster* and *C. elegans* miRNAs are now regarded as important regulatory components in many eukaryotic species including humans. The number of reported miRNAs exceeds 1,000 and many of these miRNAs have been implicated in important biological processes and human diseases.

In this volume of "Methods in Molecular Biology" we concentrate on the pathway of microRNA maturation. After its synthesis, the primary miRNA transcript (pri-miRNA) is cleaved by an endonuclease called Drosha to yield the precursor miRNA (pre-miRNA). After being transported to the cytoplasm, the latter is further cleaved by another nuclease called Dicer to yield the mature double-stranded miRNA. This mature miRNA consists of the active guide strand which remains bound to the miRNA effector complex, whereas the passenger strand is degraded. These few relatively simple steps (although the cellular machinery promoting this pathway is rather complex) are common to most miRNAs and thus are highly important to study. In this book we concentrate on three important aspects of this maturation pathway: First of all, we give an overview over the current knowledge of the pathway of miRNA maturation (Chapter 1) and how this pathway relates to human disease (Chapter 2). In the third review, established and novel approaches to manipulate miRNA maturation and activity are described.

During the last 5 years many correlations between the levels of certain miRNAs and various human malignancies have been identified, and in some cases the causative role of certain miRNAs in disease formation has been shown. Specific miRNAs can regulate certain proteins causing or preventing disease formation. These miRNA–disease correlations call for easy and reliable methods for quantifying specific miRNA species from biological samples, a topic which is only partially covered in this book. While miRNAs today have somehow lost their "exotic" touch with many standard methods established in laboratories around the world, novel methods with the potential to expand the view on miRNAs are currently being developed. Such innovation can arise from improving existing methods or by the development of completely new methods allowing for addressing questions other than previous ones.

In this book, established methods (qRT-PCR of miRNA maturation components, qRT-PCR of miRNAs) are completed with fluorescent and nonfluorescent methods for homogenous assays of Dicer-mediated miRNA maturation or an in vivo assay for Drosha activity. Since miRNAs appear to be emerging drug targets, anti-miRs are already commercially available. Less common is the use of biologically stable PNA as anti-miRs or even miRNA maturation inhibitors, doubtlessly adding to the arsenal of current oligonucleotide approaches to manipulate miRNA activity. Apart from oligonucleotides affecting miRNA activity, which are already under clinical investigation, there is a huge interest in finding small molecules with equivalent effects. Specially adapted luciferase assays have proven as powerful tools for identifying candidate small molecule miRNA effectors.

I am convinced that the collection of methods of this book is suitable to widen the view on miRNA as biological mediators and potential drug targets and thus stimulate future research in this highly dynamic and thrilling topic.

Berlin, Germany ***Christoph Arenz***

Contents

Contributors

DANILO ALLEGRA • *Internal Medicine III, University of Ulm, Ulm, Germany; German Cancer Research Center (DKFZ), Heidelberg, Germany*
CHRISTOPH ARENZ • *Institute for Chemistry, Humboldt-Universität zu Berlin, Berlin, Germany*
CONCETTA AVITABILE • *Dipartimento delle Scienze Biologiche, Università di Napoli "Federico II", Napoli, Italy*
IAN BARR • *Department of Biological Chemistry, University of California Los Angeles, Los Angeles, CA, USA*
NICOLETTA BIANCHI • *Department of Life Sciences and Biotechnology, Ferrara University, Ferrara, Italy; Laboratory for the Development of Pharmacological and Pharmacogenomic Therapy of Thalassaemia, Biotechnology Center, Ferrara University, Ferrara, Italy*
VERA BILAN • *Internal Medicine III, University of Ulm, Ulm, Germany*
ELEONORA BROGNARA • *Department of Life Sciences and Biotechnology, Ferrara University, Ferrara, Italy*
COLLEEN M. CONNELLY • *Department of Chemistry, North Carolina State University, Raleigh, NC, USA*
ROBERTO CORRADINI • *Department of Organic Chemistry, Parma University, Parma, Italy*
BRIAN P. DAVIES • *Institute for Chemistry, Humboldt Universität zu Berlin, Germany*
ALEXANDER DEITERS • *Department of Chemistry, North Carolina State University, Raleigh, NC, USA*
ENRICA FABBRI • *Department of Life Sciences and Biotechnology, Ferrara University, Ferrara, Italy*
ALESSIA FINOTTI • *Laboratory for the Development of Pharmacological and Pharmacogenomic Therapy of Thalassaemia, Biotechnology Center, Ferrara University, Ferrara, Italy*
ROBERTO GAMBARI • *Department of Life Sciences and Biotechnology, Ferrara University, Ferrara, Italy; Laboratory for the Development of Pharmacological and Pharmacogenomic Therapy of Thalassaemia, Biotechnology Center, Ferrara University, Ferrara, Italy*
FENG GUO • *Department of Biological Chemistry, University of California Los Angeles, Los Angeles, CA, USA*
STEPHAN A. HAHN • *Labor für Molekulare Gastroenterologische Onkologie (MGO), Zentrum für Klinische Forschung (ZKF), Ruhr Universität Bochum, Bochum, Germany*
ULRICH HAHN • *Institute for Biochemistry and Molecularbiology, University of Hamburg, Hamburg, Germany*
MARLEN HESSE • *Institute for Chemistry, Humboldt Universität zu Berlin, Germany*
FLORIAN KUCHENBAUER • *Internal Medicine III, University of Ulm, Ulm, Germany*
EILEEN MAGBANUA • *Institute for Biochemistry and Molecularbiology, University of Hamburg, Hamburg, Germany*

ABDELOUAHID MAGHNOUJ • *Labor für Molekulare Gastroenterologische Onkologie (MGO), Zentrum für Klinische Forschung (ZKF), Ruhr Universität Bochum, Bochum, Germany*
JACLYN R. MCKENNA • *Integrated Science Center, College of William and Mary, Williamsburg, VA, USA*
DANIEL MERTENS • *Internal Medicine III, University of Ulm, Ulm, Germany; German Cancer Research Center (DKFZ), Heidelberg, Germany*
SASKIA NEUBACHER • *Institute for Chemistry, Humboldt Universität zu Berlin, Germany*
ALESSANDRA ROMANELLI • *Dipartimento di Farmacia, Università di Napoli "Federico II", Napoli, Italy*
MICHAEL SAND • *Department of Dermatology, Venereology and Allergology, Ruhr-University Bochum, Bochum, Germany*
MARINA SKRYGAN • *Department of Dermatology, Venereology and Allergology, Ruhr-University Bochum, Bochum, Germany*
VALERIE T. TRIPP • *Integrated Science Center, College of William and Mary, Williamsburg, VA, USA*
DOUGLAS D. YOUNG • *Integrated Science Center, College of William and Mary, Williamsburg, VA, USA*
HANNAH ZÖLLNER • *Labor für Molekulare Gastroenterologische Onkologie (MGO), Zentrum für Klinische Forschung (ZKF), Ruhr Universität Bochum, Bochum, Germany*

Part I

Reviews

Chapter 1

The Pathway of miRNA Maturation

Michael Sand

Abstract

MicroRNAs (miRNAs) are a novel class of 17–23 nucleotide short, nonprotein-coding RNA molecules which have emerged to be key players in posttranscriptional gene regulation. In this chapter we give an in-depth review of the classic, canonical mammalian miRNA maturation pathway and discuss new, noncanonical alternatives such as the mirtron pathway which were recently described.

Key words MicroRNA, Primary-microRNA, Drosha, Precursor-microRNA, Dicer, Mirtron

1 Introduction: The Classic miRNA Maturation Pathway

MicroRNAs (miRNAs) are 17–23 nt short, nonprotein-coding RNA (ncRNA) molecules which are highly conserved through the genome of viruses, plants, animals, and humans. Accumulating evidence suggests their fundamental role in the concept of posttranscriptional gene regulation as they have shown to play an important role during development, cellular physiology, and disease pathology. Recently this group of ncRNAs has received considerable attention in biomedical research because of their pervasive level of gene regulation.

In this chapter we focus on the classic, canonical mammalian miRNA maturation pathway (Fig. 1) which shows a stepwise maturation pattern located inside the nucleus and in the cytoplasm and discuss new, alternative miRNA pathways recently described.

2 Intranuclear Maturation: The Primary-miRNA Transcript and Drosha Cleavage

In eukaryotes the genetic information which is necessary for the maturation of miRNAs is encoded throughout the whole genome which is stored on chromosomes inside the nucleus. In humans only approximately 1.5 % of the three billion base pairs of DNA are

Christoph Arenz (ed.), *miRNA Maturation: Methods and Protocols*, Methods in Molecular Biology, vol. 1095,
DOI 10.1007/978-1-62703-703-7_1, © Springer Science+Business Media New York 2014

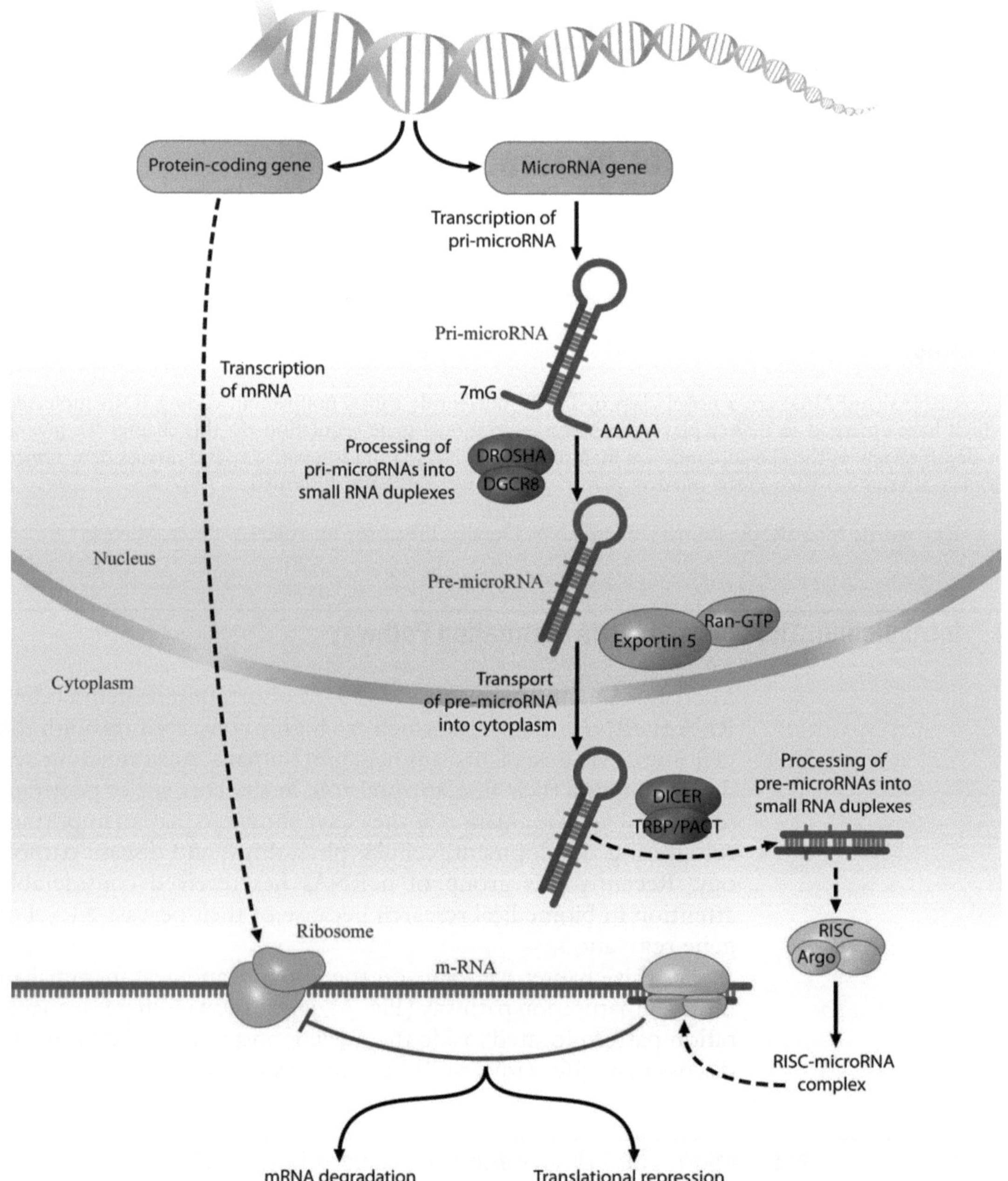

Fig. 1 The classic miRNA maturation canonical mammalian miRNA maturation pathway [1]

believed to code for proteins [1]. The primary transcript consists of protein coding and noncoding sequences. Introns located between protein-coding exons were until recently thought to be genetic wasteland consisting of meaningless code sections displaying ancient remnants of DNA that invaded our cell's DNA since the early days of evolution. Nevertheless, with the description of miRNAs this has shown to be vastly untrue. The once so-called genetic wasteland has shown to partly code for the very important primary-miRNA (pri-miRNA) transcript which is the basis for miRNAs. The (partial) highly conserved genetic information of the pri-miRNA transcripts is encoded on both exons and introns of coding and noncoding genes [2]. The intronic and exonic pri-miRNAs have been reported to be 80–87 and 13–20 % [3].

Although most miRNA sequences are encoded on introns of the host genome, some miRNAs are also encoded on exons as they show to have an overlap with transcription units [3]. The several kilobases long pri-miRNA transcript is a product of RNA polymerase II and III, whereas most of the pri-miRNAs are transcribed by RNA polymerase II inside the nucleus [4]. The pri-miRNA transcript is several kilobases long, poladenylated at its 3′ end, capped with a 7-methylguanosine cap at its 5′ end, hallmarks of polymerase II transcription, and contains a characteristic stem–loop structure [5, 6].

Depending on the genomic loci, miRNAs can be categorized into three types: intragenic (intra-miR), intergenic (inter-miR), and polycistronic (poly-miR). The evolutionarily conserved ("old") intragenic miRNAs are commonly transcribed in conjunction with their host genes, whereas it has recently been proposed that non-conserved ("young") intragenic miRNAs, which dominate in human genome, are unlikely to be co-expressed with their host genes [7, 8]. The exact mechanism of how non-conserved intragenic miRNAs are regulated is not completely understood [8]. Inter-miR are believed to be expressed independently by unidentified promoters and are located in noncoding regions [9]. Poly-miR are encoded as a cluster in the genome and are transcribed in a single primary transcript with multiple hairpins giving rise to different miRNAs [10]. After the pri-miRNA transcript has been intranuclear generated it is cleaved into several precursor-miRNAs (pre-miRNAs) by the 500–650 kDa large microprocessor complex. The core of this microprocessor complex consists of the intranuclear RNase III enzyme Drosha (RNASEN) and the double-stranded RNA-binding domain (dsRBD) protein DiGeorge syndrome critical region gene 8 (DGCR8, Pasha = *Pa*rtner of Dro*sha* in *Drosophila melanogaster* and *Caenorhabditis elegans*), which are together properly cleaving synthetic miRNA-substrates in vitro [11, 12]. In vivo however, additional accessory factors of the microprocessor complex are required for processing a subset of pri-miRNAs. DEAD-box RNA helicases p68 (DDX5) and p72 (DDX17) for example have been shown to be among those accessory factors [13–15]. DDX5 facilitates SMAD (*s*mall and *m*others *a*gainst *d*ecapentaplegic homolog) protein-mediated positive regulation of Drosha [16]. SMAD proteins are signal transducers of the transforming growth factor beta (TGFbeta) signaling pathway and can control Drosha-mediated miRNA maturation [17, 18].

Other accessory factors with influence on the microprocessor complex and pri-miRNA processing are adenosine deaminases acting on RNA (ADAR), ARS2 (a component of the nuclear RNA cap-binding complex), estrogen receptor α (ERα), Ewing sarcoma breakpoint region 1 (EWSR1), fused in sarcoma (Fus), KH-type splicing regulatory protein (KHRSP), numerous heterogenous nuclear RNA complex (hnRNP) proteins, nuclear factor 90 and nuclear factor 45 complex (NF90–NF45 complex), TAR DNA-binding protein-43 (TDP-43), cardinal tumor suppressor protein 53 (p53), and 396-amino acid long SMAD nuclear interacting protein 1 (SNIP1) [19–27]. ADAR for example has a prominent role in miRNA editing of a subset of pri-miRNA transcripts as it modifies adenosine (A) to inosine (I) by deamination. Inosine shares similar base properties with guanine (G). A-to-I editing can change the actual miRNA precursor sequence and therefore its base-pairing, structural, and target recognition properties [28]. A-to-I editing has been described for miR-22, miR-151, miR-197, miR-223, miR-376a, miR-379, miR-99a, miR-142, miR-223, miR-1-1, and miR-143 [19, 29, 30].

Drosha cofactor DGCR8 has two double-stranded RNA (dsRNA)-binding sites and is pivotal for Drosha-mediated cutting of the pri-miRNA transcript as it specifically facilitates binding to the pri-miRNA cutting site of Drosha [31]. The pri-miRNA contains a dsRNA hairpin stem of ~33 base pairs and a terminal loop with imperfect complementarity and flanking single-stranded RNA (ssRNA) at its base. DGCR-8 recognizes the ssRNA/dsRNA junction of the pri-miRNA. The transcription of the pri-miRNA transcript is followed by the generation of pre-miRNAs. The pri-miRNA is cut by the two RNase domains of Drosha at the ssRNA/dsRNA junction, about 11 nt away from the pri-miRNA hairpin base resulting in a pre-miRNA with a small 3′ overhang as a result of asymmetrical cleavage of dsRNA by the microprocessor complex [32, 33]. Homeostatic control of miRNA biogenesis is achieved through crossregulation between Drosha and DGCR8. DGCR8 stabilizes the Drosha protein via protein–protein interaction between its conserved carboxy-terminal domain and the middle domain of Drosha [34]. Drosha in turn cleaves two hairpin structures embedded in the DGCR8 mRNA [35]. Cleavage of the two hairpins in the DGCR8 mRNA, one in the 5′ untranslated region (UTR) and the other in the coding region close to the start of the open reading frame (ORF), leads to its destabilization and decreased DGCR8 expression [33, 36]. This posttranscriptional regulation of DGCR8 by the microprocessor is an autoregulatory, negative feedback mechanism controlling the microprocessor. Interestingly deep-sequencing analyses have shown that DGCR8 mRNA is the only mRNA which is targeted by the microprocessor [37].

The resulting product of the first microprocessor-mediated cutting event is called pre-miRNA, is 60–100 nts in length, and has a characteristic two-nucleotide overhang at the 3′ terminal end.

This 2 nt 3′-overhang of the pre-miRNA is bound by exportin-5 (Exp-5, Ran-binding protein 21), member of the karyopherin family of transport proteins and part of a 230 kDa protein complex which not only protects it from cellular digestion but also mediates pre-miRNA translocation from the nucleus to the cytoplasm in an energy-reliant process [38–40]. During this translocation process Exp-5 interacts with Ras-related nuclear protein (Ran), a small 25-kDa protein involved in transport in and out of the cell nucleus. In order that the pre-miRNA is bound by the Exp-5/Ran complex, Ran needs to be in the Ran-guanine triphosphatase (Ran-GTP) state. The pre-miRNA/Exp-5/Ran-GTP complex is then translocated to the cytoplasm through a nuclear pore complex. Once in the cytoplasm, interaction with the Ran-GTPase-activating protein (RanGAP) facilitates Ran-GTPase activity [41]. In consequence of Ran-GTPase activity Exp-5-bound Ran-GTP is hydrolized to Ran-guanine diphosphatase (Ran-GDP). The Ran-GTP hydrolysis leads to pre-miRNA release in the cytoplasm [11].

3 Extranuclear Maturation: The Cytosolic Precursor-miRNAs and Dicer Cleavage of Pre-miRNAs to the Mature miRNA

Once the pre-miRNA has been exported to the cytoplasm a second-round ribonuclease reaction takes place: the RNase III enzyme Dicer together with its cofactors dsRBD protein Tar RNA-binding protein (TRBP) and protein activator of PKR (PACT) cleaves the pre-miRNA into an ~22 nt miRNA/miRNA* duplex with a 2 nt overhang at each of its 3′ end by removing the pre-miRNA hairpin loop [42]. One strand, the mature miRNA guide strand, is loaded onto Argonaute2 (AGO2) which forms the functional center of an RNA-induced silencing complex (RISC) [43]. The mature miRNA guide strand guides RISC, the cytoplasmatic effector molecule in RNA interference (RNAi), to the mRNA target sequence. In humans, Dicer cleavage activity is regulated via the amino-terminal DExD/H-box helicase domain which inhibits Dicer activity. TRBP activates Dicer by binding to this region which results in conformational rearrangement [44]. The Dicer product, miRNA let-7, targets Dicer mRNA, establishing a miRNA/Dicer autoregulatory negative feedback loop [45]. The miRNA–maturing enzyme Dicer, the dsRBD TRBP, PACT, and Ago2 form a RISC-loading complex (RLC) which facilitates transfer of mature miRNAs from Dicer to RISC. In some cases when there is a high degree of complementarity in the hairpin stem of the pre-miRNA, Ago2 generates an additional processing intermediate, termed Ago2-cleaved precursor miRNA or ac-pre-miRNA [46]. This ac-pre-miRNA is nicked at the miRNA* passenger strand as a result of Ago2-mediated cleavage of the 3′ arm of the pre-miRNA hairpin. Dicer is able to further process the ac-pre-miRNA as other regular pre-miRNAs [28]. Regarding the selection of the miRNA guide strand and its counterpart the miRNA* passenger strand, it has been shown that the RNA strand which is less thermodynamically stable at the 5′end is designated as the miRNA guide strand whereas the remaining strand will become the miRNA* passenger strand which is usually discarded [11, 47, 48]. Nevertheless miRNA* strands show regulatory function in vertebrates and have Ago sorting characteristics in Drosophila [49–55]. Once bound, dsRNAs undergo significant repositioning within the Dicer complex following dicing which enables directional binding of RNA duplexes. There are two distinct RNA-binding sites within the Dicer complex—one for dsRNA processing and the other for sensing thermodynamic asymmetry which facilitates guide strand selection during RISC loading in humans [56]. Once the mature miRNA guide strand has been loaded on the RISC, the miRNA/RISC complex is ready to anneal to the target mRNA in the so-called miRNA seed region on nts 2–7. In the event of nearly complete complementarity the target mRNA is cleaved endonucleolytically resulting in a cessation of translation, while translation is only suppressed in case of limited complementarity [57, 58].

4 Mirtrons: The First Alternative miRNA Maturation Pathway

After the canonical miRNA pathway had been specified as previously illustrated a growing number of noncanonical miRNA pathways have been described [55, 59]. The first and most important alternative miRNA maturation pathway described is the "mirtron" pathway.

Mirtrons are directly spliced out from host genes as pre-miRNAs with hairpin potential and do not depend on pri-miRNA Drosha/DGCR8 processing. The product of intron splicing is not linear but mimics a transient lariat shape wherein the 3′ branchpoint is ligated to the 5′ end of the intron. After the lariat-shape intron is processed by the lariat debranching enzyme (Ldbr) it refolds into a pre-miRNA shape which is recognized and transferred to the cytoplasm by Exp-5. Following regular Dicer cleavage the mature miRNA can be loaded on the RISC and exerts its effect.

References

1. Katayama S, Tomaru Y, Kasukawa T, Waki K, Nakanishi M, Nakamura M, Nishida H, Yap CC, Suzuki M, Kawai J, Suzuki H, Carninci P, Hayashizaki Y, Wells C, Frith M, Ravasi T, Pang KC, Hallinan J, Mattick J, Hume DA, Lipovich L, Batalov S, Engstrom PG, Mizuno Y, Faghihi MA, Sandelin A, Chalk AM, Mottagui-Tabar S, Liang Z, Lenhard B, Wahlestedt C (2005) Antisense transcription in the mammalian transcriptome. Science 309:1564–1566
2. Kim YK, Kim VN (2007) Processing of intronic microRNAs. EMBO J 26:775–783
3. Rodriguez A, Griffiths-Jones S, Ashurst JL, Bradley A (2004) Identification of mammalian microRNA host genes and transcription units. Genome Res 14:1902–1910
4. Borchert GM, Lanier W, Davidson BL (2006) RNA polymerase III transcribes human microRNAs. Nat Struct Mol Biol 13: 1097–1101

5. Lee Y, Jeon K, Lee JT, Kim S, Kim VN (2002) MicroRNA maturation: stepwise processing and subcellular localization. Embo J 21: 4663–4670
6. Sand M, Gambichler T, Sand D, Skrygan M, Altmeyer P, Bechara FG (2009) MicroRNAs and the skin: tiny players in the body's largest organ. J Dermatol Sci 53:169–175
7. Bhattacharyya M, Feuerbach L, Bhadra T, Lengauer T, Bandyopadhyay S (2012) MicroRNA Transcription Start Site Prediction with Multi-objective Feature Selection. Stat Appl Genet Mol Biol 11:1–25
8. He C, Li Z, Chen P, Huang H, Hurst LD, Chen J (2012) Young intragenic miRNAs are less coexpressed with host genes than old ones: implications of miRNA-host gene coevolution. Nucleic Acids Res 40:4002–4012
9. Li SC, Tang P, Lin WC (2007) Intronic microRNA: discovery and biological implications. DNA Cell Biol 26:195–207
10. Marco A, Hooks K, Griffiths-Jones S (2012) Evolution and function of the extended miR-2 microRNA family. RNA Biol 9
11. Lau PW, MacRae IJ (2009) The molecular machines that mediate microRNA maturation. J Cell Mol Med 13:54–60
12. Gregory RI, Yan KP, Amuthan G, Chendrimada T, Doratotaj B, Cooch N, Shiekhattar R (2004) The Microprocessor complex mediates the genesis of microRNAs. Nature 432:235–240
13. Lee Y, Ahn C, Han J, Choi H, Kim J, Yim J, Lee J, Provost P, Radmark O, Kim S, Kim VN (2003) The nuclear RNase III Drosha initiates microRNA processing. Nature 425:415–419
14. Trabucchi M, Briata P, Filipowicz W, Rosenfeld MG, Ramos A, Gherzi R (2009) How to control miRNA maturation? RNA Biol 6:536–540
15. Fukuda T, Yamagata K, Fujiyama S, Matsumoto T, Koshida I, Yoshimura K, Mihara M, Naitou M, Endoh H, Nakamura T, Akimoto C, Yamamoto Y, Katagiri T, Foulds C, Takezawa S, Kitagawa H, Takeyama K, O'Malley BW, Kato S (2007) DEAD-box RNA helicase subunits of the Drosha complex are required for processing of rRNA and a subset of microRNAs. Nat Cell Biol 9:604–611
16. Davis BN, Hilyard AC, Lagna G, Hata A (2008) SMAD proteins control DROSHA-mediated microRNA maturation. Nature 454:56–61
17. Hata A, Davis BN (2010) Regulation of pri-miRNA processing through Smads. Adv Exp Med Biol 700:15–27
18. Davis BN, Hilyard AC, Nguyen PH, Lagna G, Hata A (2010) Smad proteins bind a conserved RNA sequence to promote microRNA maturation by Drosha. Mol Cell 39:373–384
19. Yang W, Chendrimada TP, Wang Q, Higuchi M, Seeburg PH, Shiekhattar R, Nishikura K (2006) Modulation of microRNA processing and expression through RNA editing by ADAR deaminases. Nat Struct Mol Biol 13:13–21
20. Gruber JJ, Zatechka DS, Sabin LR, Yong J, Lum JJ, Kong M, Zong WX, Zhang Z, Lau CK, Rawlings J, Cherry S, Ihle JN, Dreyfuss G, Thompson CB (2009) Ars2 links the nuclear cap-binding complex to RNA interference and cell proliferation. Cell 138:328–339
21. Yamagata K, Fujiyama S, Ito S, Ueda T, Murata T, Naitou M, Takeyama K, Minami Y, O'Malley BW, Kato S (2009) Maturation of microRNA is hormonally regulated by a nuclear receptor. Mol Cell 36:340–347
22. Trabucchi M, Briata P, Garcia-Mayoral M, Haase AD, Filipowicz W, Ramos A, Gherzi R, Rosenfeld MG (2009) The RNA-binding protein KSRP promotes the biogenesis of a subset of microRNAs. Nature 459:1010–1014
23. Trabucchi M, Briata P, Filipowicz W, Ramos A, Gherzi R, Rosenfeld MG (2011) KSRP promotes the maturation of a group of miRNA precuresors. Adv Exp Med Biol 700:36–42
24. Sakamoto S, Aoki K, Higuchi T, Todaka H, Morisawa K, Tamaki N, Hatano E, Fukushima A, Taniguchi T, Agata Y (2009) The NF90-NF45 complex functions as a negative regulator in the microRNA processing pathway. Mol Cell Biol 29:3754–3769
25. Suzuki HI, Yamagata K, Sugimoto K, Iwamoto T, Kato S, Miyazono K (2009) Modulation of microRNA processing by p53. Nature 460: 529–533
26. Yu B, Bi L, Zheng B, Ji L, Chevalier D, Agarwal M, Ramachandran V, Li W, Lagrange T, Walker JC, Chen X (2008) The FHA domain proteins DAWDLE in Arabidopsis and SNIP1 in humans act in small RNA biogenesis. Proc Natl Acad Sci U S A 105:10073–10078
27. Kawahara Y, Mieda-Sato A (2012) TDP-43 promotes microRNA biogenesis as a component of the Drosha and Dicer complexes. Proc Natl Acad Sci U S A 109:3347–3352
28. Winter J, Jung S, Keller S, Gregory RI, Diederichs S (2009) Many roads to maturity: microRNA biogenesis pathways and their regulation. Nat Cell Biol 11:228–234
29. Luciano DJ, Mirsky H, Vendetti NJ, Maas S (2004) RNA editing of a miRNA precursor. RNA 10:1174–1177
30. Blow MJ, Grocock RJ, van Dongen S, Enright AJ, Dicks E, Futreal PA, Wooster R, Stratton MR (2006) RNA editing of human microRNAs. Genome Biol 7:R27
31. Landthaler M, Yalcin A, Tuschl T (2004) The human DiGeorge syndrome critical region gene 8 and Its D. melanogaster homolog are required for miRNA biogenesis. Curr Biol 14:2162–2167
32. Han J, Lee Y, Yeom KH, Kim YK, Jin H, Kim VN (2004) The Drosha-DGCR8 complex in

primary microRNA processing. Genes Dev 18:3016–3027
33. Triboulet R, Gregory RI (2010) Autoregulatory mechanisms controlling the Microprocessor. Adv Exp Med Biol 700:56–66
34. Yeom KH, Lee Y, Han J, Suh MR, Kim VN (2006) Characterization of DGCR8/Pasha, the essential cofactor for Drosha in primary miRNA processing. Nucleic Acids Res 34: 4622–4629
35. Han J, Pedersen JS, Kwon SC, Belair CD, Kim YK, Yeom KH, Yang WY, Haussler D, Blelloch R, Kim VN (2009) Posttranscriptional cross-regulation between Drosha and DGCR8. Cell 136:75–84
36. Pedersen JS, Bejerano G, Siepel A, Rosenbloom K, Lindblad-Toh K, Lander ES, Kent J, Miller W, Haussler D (2006) Identification and classification of conserved RNA secondary structures in the human genome. PLoS Comput Biol 2:e33
37. Shenoy A, Blelloch R (2009) Genomic analysis suggests that mRNA destabilization by the microprocessor is specialized for the auto-regulation of Dgcr8. PLoS One 4:e6971
38. Guttler T, Gorlich D (2011) Ran-dependent nuclear export mediators: a structural perspective. EMBO J 30:3457–3474
39. Bohnsack MT, Czaplinski K, Gorlich D (2004) Exportin 5 is a RanGTP-dependent dsRNA-binding protein that mediates nuclear export of pre-miRNAs. RNA 10:185–191
40. Lund E, Guttinger S, Calado A, Dahlberg JE, Kutay U (2004) Nuclear export of microRNA precursors. Science 303:95–98
41. Coutavas E, Ren M, Oppenheim JD, D'Eustachio P, Rush MG (1993) Characterization of proteins that interact with the cell-cycle regulatory protein Ran/TC4. Nature 366:585–587
42. Sand M, Skrygan M, Georgas D, Arenz C, Gambichler T, Sand D, Altmeyer P, Bechara FG (2011) Expression levels of the microRNA maturing microprocessor complex component DGCR8 and the RNA-induced silencing complex (RISC) components Argonaute-1, Argonaute-2, PACT, TARBP1, and TARBP2 in epithelial skin cancer. Mol Carcinog 51: 916–922
43. Wang HW, Noland C, Siridechadilok B, Taylor DW, Ma E, Felderer K, Doudna JA, Nogales E (2009) Structural insights into RNA processing by the human RISC-loading complex. Nat Struct Mol Biol 16:1148–1153
44. Ma E, MacRae IJ, Kirsch JF, Doudna JA (2008) Autoinhibition of human dicer by its internal helicase domain. J Mol Biol 380:237–243
45. Forman JJ, Legesse-Miller A, Coller HA (2008) A search for conserved sequences in coding regions reveals that the let-7 microRNA targets Dicer within its coding sequence. Proc Natl Acad Sci U S A 105:14879–14884
46. Diederichs S, Haber DA (2007) Dual role for argonautes in microRNA processing and post-transcriptional regulation of microRNA expression. Cell 131:1097–1108
47. Schwarz DS, Hutvagner G, Du T, Xu Z, Aronin N, Zamore PD (2003) Asymmetry in the assembly of the RNAi enzyme complex. Cell 115:199–208
48. Khvorova A, Reynolds A, Jayasena SD (2003) Functional siRNAs and miRNAs exhibit strand bias. Cell 115:209–216
49. Okamura K, Phillips MD, Tyler DM, Duan H, Chou YT, Lai EC (2008) The regulatory activity of microRNA* species has substantial influence on microRNA and 3′ UTR evolution. Nat Struct Mol Biol 15:354–363
50. Ro S, Park C, Young D, Sanders KM, Yan W (2007) Tissue-dependent paired expression of miRNAs. Nucleic Acids Res 35:5944–5953
51. Yang JS, Phillips MD, Betel D, Mu P, Ventura A, Siepel AC, Chen KC, Lai EC (2011) Widespread regulatory activity of vertebrate microRNA* species. RNA 17:312–326
52. Czech B, Zhou R, Erlich Y, Brennecke J, Binari R, Villalta C, Gordon A, Perrimon N, Hannon GJ (2009) Hierarchical rules for Argonaute loading in Drosophila. Mol Cell 36:445–456
53. Ghildiyal M, Xu J, Seitz H, Weng Z, Zamore PD (2010) Sorting of Drosophila small silencing RNAs partitions microRNA* strands into the RNA interference pathway. RNA 16:43–56
54. Okamura K, Liu N, Lai EC (2009) Distinct mechanisms for microRNA strand selection by Drosophila Argonautes. Mol Cell 36:431–444
55. Westholm JO, Lai EC (2011) Mirtrons: microRNA biogenesis via splicing. Biochimie 93:1897–1904
56. Noland CL, Ma E, Doudna JA (2011) siRNA repositioning for guide strand selection by human Dicer complexes. Mol Cell 43:110–121
57. Reinhart BJ, Slack FJ, Basson M, Pasquinelli AE, Bettinger JC, Rougvie AE, Horvitz HR, Ruvkun G (2000) The 21-nucleotide let-7 RNA regulates developmental timing in Caenorhabditis elegans. Nature 403:901–906
58. Dalmay T (2008) Identification of genes targeted by microRNAs. Biochem Soc Trans 36: 1194–1196
59. Miyoshi K, Miyoshi T, Siomi H (2010) Many ways to generate microRNA-like small RNAs: non-canonical pathways for microRNA production. Mol Genet Genomics 284:95–103

Chapter 2

MicroRNA Maturation and Human Disease

Marlen Hesse and Christoph Arenz

Abstract

Numerous studies describe alterations in the levels of specific microRNAs (miRNAs) that are associated with human pathologies. Some of these alterations may give rise to the development of novel diagnostic tools, while certain miRNAs additionally could serve as novel drug targets. Moreover, components of the miRNA maturation machinery may be up- or down-regulated in human disease. In such cases, the consequences for the expression of individual miRNAs are however only poorly understood. Herein, we review the current knowledge of how miRNAs are linked to human disease and which parts of the miRNA maturation machinery could serve as future drug targets.

Key words miRNAs, Dicer, Drosha, Biomarkers, Drug targets

1 miRNA Function and Biogenesis

MicroRNAs (miRNAs) are small, noncoding, 21–25 nt regulatory RNA molecules involved in physiological and pathological processes [1, 2]. First footsteps in discovering miRNA function and its relevancy in gene regulation were set by identifying miRNA family *lin-4* in *Caenorhabditis elegans* [3]. Lin-4 is indispensable for the normal temporal control of postembryonic developmental events in *C. elegans* organism. Lee et al. found that lin-4 does not only *not* encode a protein, but, more importantly, it regulates lin-14 translation through "RNA-RNA interaction". By negatively regulating lin-14 protein levels, lin-4 enables a temporal decrease of that protein during the first larval stage (L1). This interaction was confirmed by identifying two small lin-4 transcripts of about 20 and 60 nt containing sequences complementary to a 3′-untranslated sequence in lin-14 mRNA. Lin-14 protein is down-regulated during late L1 and introduces switching to second larval stage (L2). In fact, mutations in lin-14 let to unusual high amounts of lin-14 protein in nucleus from the beginning of L2 up to adult stages [4]. Analysis of those mutations clarified that they are lesions in the 3′-untranslated region (UTR) of lin-14 [5]. As no protein-coding

Christoph Arenz (ed.), *miRNA Maturation: Methods and Protocols*, Methods in Molecular Biology, vol. 1095, DOI 10.1007/978-1-62703-703-7_2, © Springer Science+Business Media New York 2014

regions are involved, mutations do not alter or destabilize the protein itself, but most likely interfere with posttranscriptional processes. Finally, demonstration that the mysterious "RNA–RNA interaction" equals complementary binding of miRNA lin-4 to lin-14 mRNA suggested a whole new noncoding RNA-world to be investigated [6, 7]. The second miRNA discovered as recently as in the year 2000 is let-7 encoding a 21 nt small RNA also complementary to elements in the 3′-UTR of lin-14 among others, like lin-28 and lin-41 [8]. In *C. elegans* let-7 functions as a heterochromic switch gene between larval and adult stage coordinating developmental timing. Surprisingly, both let-7 and lin-41 are evolutionarily conserved with homologues readily detected in mollusks, flies, mice, and humans [9]. Database searches (BLASTN) for matches to let-7 RNA reveal DNA segments from *Drosophila melanogaster* as well as from the human genome sequence. As for precursor transcripts of let-7 RNA, similar stem–loop secondary structures are predicted for *Caenorhabditae*, *Drosophila*, and *H. sapiens*. Interestingly, only this similar RNA fragment can be detected in most species, emphasizing that let-7 may be efficiently processed from a putative precursor to a mature form of miRNA. Raising proof indicates that let-7 is essential for gene regulation in late temporal transitions across animal phylogeny.

Known by now, mature miRNA mostly derives from one arm of larger precursor miRNA which forms imperfect stem–loop structure [10]. Biogenesis of miRNA starts within the nucleus: miRNAs are encoded in the genome as pri-miRNA [11]. These long primary transcripts extend both 5′ and 3′ from the miRNA sequence and contain a cap structure at the 5′-end and a polyadenylated 3′-end. Primary transcript is successively processed by two ribonucleases (RNase III types) to its mature form [12]. Two sequential reactions trim the transcript into mature miRNA. Processing basically depends on the formation of miRNA into a stem–loop structure [10]. First, pri-miRNAs are processed into ~70 nt precursors (pre-miRNA) by cellular multiprotein complex predominantly located in the nucleus. Its core components are RNase III enzyme Drosha and double-stranded RNA-binding domain DGCR8/Pasha. Drosha cleaves cotranscriptionally generating pre-miRNA with an RNase III characteristic 2 nt 3′-overhang. Efficiency of Drosha processing depends on pre-miRNA's terminal loop size, stem structure, and flanking sequence of the Drosha cleavage site [13]. The following export from the nucleus into the cytoplasm by exportin 5 (Exp5) is initiated by 3′-overhang recognition. Inside the cytoplasm, the hairpin pre-miRNA is cleaved by second RNase III enzyme Dicer into a small, imperfect dsRNA duplex. Double-stranded RNA now consists of both mature miRNA strand and its complementary strand ranging from 21 to 25 nucleotides. Dicer enzyme is characterized by several domains in specific order. From the amino-to-carboxy terminus it contains a putative helicase domain, a DUF283 domain, a Piwi–Argonaute–Zwille (PAZ) domain, two tandem RNase III domains, and a dsRNA-binding domain. PAZ domain interacts with dsRNAs that present the 2 nt 3′-overhangs and a defined minimal stem length probably necessary for recognizing the end of Drosha cleavage product and for enabling Dicer cleavage. Processing occurs at dsRNA ends that associate with the PAZ domain. Then, about two helical turns along the surface of the protein, the substrate extends to a single processing centre followed by RNase III domains and each of the two RNase III domains cleaves one of the two strands. The reaction leaves new ends with ~2 nt 3′-overhangs and a 5′-monophosphate on the product ends required during later stages of silencing [14]. The double-stranded products of Dicer can be incorporated into RNA-induced silencing complex (RISC) to induce posttranscriptional gene-silencing mechanisms. Argonaute (Ago) proteins are the central components of all RISCs. The Ago clade associates with miRNA and siRNA during the RISC assembly pathway, which involves duplex unwinding and culminating in the stable association of only one of the two strands with Ago. The associated guide strand leads in target recognition by Watson–Crick base pairing, whereas the passenger strand of the original RNA duplex is discarded.

Functioning as translational repressors, most miRNAs in animals are only able to bind to the 3′-UTR with imperfect complementarity [10]. Despite the mRNA translational repression model, there sure are miRNAs that bind perfectly complementary to its corresponding mRNA, which then leads to mRNA cleavage. Another comparable posttranscriptional gene-silencing mechanism that is evolutionary conserved and sequence specific is called RNA interference (RNAi) [15]. It is derived by processing dsRNA to small interfering RNAs (siRNAs) initiating mRNA cleavage by perfect complementary binding.

Concerning molecular characteristics, biogenesis, and effector functions, miRNAs and siRNAs are similar. Both RNA types share Dicer as RNase-III processing enzyme and the strongly related effector complex RISC for posttranscriptional repression. The difference between miRNA and siRNA is its molecular origin and the mode of mRNA target recognition. miRNA emerges from specific precursors encoded in the genome into distinct species, whereas siRNA is sampled almost randomly from longer dsRNA [10, 15]. That dsRNA can be induced exogenously or produced from endogenously transcribed and annealed RNA strands. By contrast to the often imperfect binding of miRNA to its target, siRNA mostly forms perfect duplexes with target mRNA. Consequently, the distinction between miRNA and siRNA is open for discussion and may be elucidated by further investigations.

Over the recent years, interest in finding ways to manipulate miRNA function increased steadily not only to elucidate mechanisms of miRNA function but also with the perspective to affect human disease formation and progression [16]. As the expression pattern of certain miRNAs correlates with a variety of human diseases like cancer, effort is directed towards acquisition of simple and reliable methods to map miRNA expression profiles. In order to specifically recognize the "signature" of for example cancerogenesis, several approaches to control miRNA function by manipulating its formation or its maturation have been described [17].

2 miRNAs in Human Disease

Given the importance of miRNAs for gene regulation especially in the context of developmental processes and keeping in mind the high number of ~1,500 known human miRNAs, the discovery of possible roles for miRNAs in human pathology is not surprising. Early research into that direction was directed towards the expression levels of miRNAs in diseases characterized by a lack of diagnostic tools, especially in cancer. For example, metastatic cancer of unknown primary site (CUP) is characterized by lack of primary tumors. Lu et al. analyzed 17 poorly differentiated tumors by using a training set of about 70 more differentiated tumors belonging to 11 different types as a classifier based on available miRNA profiles [18]. Surprisingly, the accuracy of diagnoses based on 217 miRNAs was higher as compared to a set of 16,000 mRNAs. This and similar studies initiated a tremendous effort in the investigation of the role of miRNA in human disease. In general, it can be stated that there are different kinds of links between miRNA and human disease, which include (a) miRNAs of viral origins, (b) miRNAs as biomarkers that only reflect disease states, (c) miRNAs with individual roles as triggers or suppressors of disease development, and (d) alterations in the miRNA processing machinery that may affect the expression of many if not all miRNAs. Lastly, we will point out novel lines of research aimed at gaining control over miRNA-mediated disease.

3 Viral miRNAs

Some disease-related miRNAs are not of human origin but are encoded by viruses [19]. Virus-encoded miRNAs have been shown to directly downregulate factors of the host's immune system [20]. In addition, viruses have also evolved the ability to downregulate or upregulate the expression of specific cellular miRNAs to enhance their replication. Some viruses gain control over cellular processes by mimicking the host's miRNAs [21], while other virus-encoded miRNAs are directed against the viral mRNAs [19]. The significance of viral miRNAs is impressively highlighted by the fact that viral miRNAs can be critical for host cell transformation in vivo [22].

Like in eukaryote species, viral miRNAs stem from larger miRNA precursors which are cleaved by the host's miRNA processing machinery [20]. So far, miRNAs have only been detected in DNA viruses, probably since a Drosha-mediated excision of a pri-miRNA would lead to a cleavage of an RNA-genome.

In contrast, viruses not only use their own miRNAs but can also interact with miRNAs of the host. Accordingly, it has been shown that hepatitis C virus (HCV) replication essentially depends on a host's cellular miRNA [23], which makes liver-specific miR-122 a potential drug target against HCV infections (see below). Although human miR-122 is not a disease-causing factor, it is essential in HCV infection and thus appears to be a promising drug target.

4 Diagnostic Tool: miRNA Profiling

Numerous reports describe association between altered miRNA expression levels and human disease, irrespective of whether these alterations are the cause for disease formation or just reflect a pathological condition. Despite the urge to find the actual active molecules that are targeted by mature miRNAs, recent findings emphasize that miRNA expression profiling can already serve as a diagnostic tool and this is at least the starting point for good cancer therapy. Indeed, miRNA levels match certain expression patterns individually composed for many different types of cancer. Therefore, miRNA profiling as a diagnostic and prognostic tool has become valuable especially in diagnosing tumor types that cannot be identified with biopsy samples on the basis of histological diagnosis alone. Metastatic CUP is characterized by lack of primary tumor. In chronic lymphocytic leukemia miRNA expression signature of 13 genes is associated with prognostic factor ZAP-70, a predictor of early disease progression [24]. Expression signature differentiates between leukemia with low and high ZAP-70 expression levels and allows the prediction whether there will be disease progression or not. miRNAs may be useful prognostic and diagnostic markers in lung adenocarcinomas, because miRNA signatures differ with tumor histology [25]. Additionally, high miR-155 and low miR-let-7a expression showed correlation with poor outcome. Another example of miRNA as a prognostic marker is the correlation of miRNA expression with specific pathologic factors in breast cancer [26]. Among 29 differentially expressed miRNAs, miR-125b, miR-145, miR-21, and miR-155 were most consistently deregulated in cancer tissue. Comparison of various pathologic features, such as estrogen and progesterone receptor expression and tumor stage, with miRNA expression levels indicated a small number of differentially expressed miRNAs according to specific breast cancer type. Many more correlations between aberrant miRNA profiles and human disease exist and have been extensively reviewed [27, 28].

Very recently, the so-called circulating miRNAs gained special attention [29]. These miRNAs are found in blood and other body fluids, in which they are obviously resistant to rapid degradation [30]. Circulating miRNAs are especially attractive as candidates for novel biomarkers, since they can be easily obtained without the need for biopsies and other invasive operations [31]. There are many reports on significant associations of circulating miRNA patterns to certain disease states [29]. Since the mechanisms by which circulating miRNAs are released to body fluids are not fully understood, it is not clear whether such miRNA species would also be potential drug targets. It seems clear that circulating miRNAs are not only remnants of dying cells, but they are also usually packaged into lipid microvesicles like exosomes [32] or apoptotic bodies [33]. Lipid vesicles not only contain distinct patterns of miRNAs, suggesting a controlled mechanism of packaging which is dependent on neutral sphingomyelinase-2 [34]. In addition, it has been described that such wrapped miRNAs can affect regulatory processes in other cells [32]. Vesicle-free miRNAs either bound to HDL [35] or to proteins that are part of the miRISC like Argonaute 2 [36] are also known. Undoubtedly, the existence of circulating miRNAs and their tightly regulated delivery to body fluids add another level of complexity to the question, why miRNAs are often aberrantly expressed in the context of human pathology.

Although regulation of miRNA biogenesis is an important matter, it has not been studied extensively enough yet. Surprisingly, evidence rises that many miRNAs are controlled by the very targets they regulate. Posttranscriptional regulation of miRNA biogenesis in *C. elegans* is realized by a double-negative feedback loop: Translation of Lin28 is inhibited by let7 miRNA, which turns out to be inhibited by Lin28 protein [37]. This tightly regulated mechanism of action also provides hints why miRNA is largely involved into cancer initiation and progression. As a matter of fact, miRNA expression profiles are useful for classification, diagnosis, and prognosis of human malignancies. Participation in essential biological processes is proved for several miRNAs, among them cell proliferation control, hematopoietic B-cell lineage fate, B-cell survival, brain patterning, pancreatic cell insulin secretion, and adipocyte development [38]. Of course, to gather the facts about tumor suppressor and oncogenic miRNAs an armada of tools was and is going to be developed to characterize cancer genomics as precise as possible. miRNA microarrays [39] developed for high-throughput miRNA fingerprint investigation in cells was the first step achievement followed by other technologies: macroarrays [40], bead-based flow cytrometric miRNA expression [18], and quantitative reverse transcription PCR [41]. Many more techniques for a multiplexed and quantitative detection of miRNAs are currently under development [42]. Besides the basic technologic concerns with miRNA detection, a further important objective will be the definition of standards for miRNA normalization in order to allow for inter-study comparisons [43]. Usually, changes in miRNA expression are normalized to other transcripts [44]. This can sometimes be error-prone due to inhomogeneities in miRNA ends. In contrast, in miRNA quantification by means of deep sequencing the relative occurrence of an individual sequence is adequately monitored [45]. A further step towards this direction might be done by absolute quantification of miRNAs [46].

5 miRNAs Directly Influencing Human Disease

Although the cause of miRNA misexpression in cancer is not quite clear, it is known that many miRNAs reside in genomic regions involved in cancer [38]. miRNAs appear to be able to contribute to oncogenesis either as tumor suppressors (miR-15a and miR-16-1) or oncogenes (mir-17-92 cluster, mir-122). Reasons of abnormal expression of miRNA genes in malignant cells are mutations, genomic amplifications or deletions, and chromosomal rearrangement—just as for protein-coding genes [47]. And because an apparently high number of miRNA genes is located in cancer-associated genomic regions, an aberrant expression causing cancer is thought to be likely. One of the most common abnormalities in human cancer is the dysregulated expression of c-Myc [48]. The oncogene c-Myc encodes a transcription factor that regulates cell proliferation and survival via several targets including E2F1 transcription factor [49]. C-Myc activates E2F1 but also expression of a six-membered miRNA cluster on human chromosome 13 (mir-17-92 cluster). Two miRNAs among them negatively regulate expression of E2F1, which leads to a strongly controlled proliferative signal. To emphasize, c-Myc is able to stimulate E2F1 transcription and to withhold its translation via miRNA activation. Tumor development studies in a mouse B-cell lymphoma model indicate that miRNAs are able to modulate tumor formation implicating the mir-17-92 cluster as potential human oncogene [50]. Human tumor samples of different types of lymphoma reveal significant overexpression of pri-mir-17-92 in 65 % of the samples, whereas colorectal carcinomas showed less overexpression from this locus. Although it is suggested that miRNA overproduction from that cluster weakens the proapoptotic response to elevated Myc expression in vivo, it remains unclear how it is promoted exactly. Lack of knowledge is also blamable on the lack of biochemical strategies allowing to identify the *targets* of miRNAs. More than 50 % of miRNA genes are found within regions connected to amplification, deletion, and translocation in cancer, even though miRNAs represent only 1 % in the mammalian genome [51]. Changes in miRNA function seem to be generated mostly by abnormal gene expression pattern that is then characterized by deviant expression levels of mature or precursor miRNA or both of them. This means that abnormally expressed miRNAs affect transcripts of essential protein-coding genes relevant in tumorigenesis, just as E2F1 transcription factor is targeted by the mir-17-92 cluster.

One theory focuses on the assumption that miRNAs expressed in various forms of cancer are from the same pool of miRNAs, differentially expressed compared to corresponding normal tissue but with different signatures for each tumor or cancer type [52]. Volinia et al. compared six solid cancers and found miR-21 to be overexpressed in all of them. Two others, miR-17-5p and miR-191, were overexpressed in five cancer types. The answer of why same miRNAs occur abnormally expressed in tumors with different embryological origin may be a shared signaling pathway altered in many kinds of tumors. MiR-21, for example, is thought to directly target PTEN, a tumor suppressor encoding a phosphatase that inhibits growth or survival pathways, which suggests that miR-21 is an antiapoptotic factor in different tumor types [53].

In case of miR-17-92 locus, investigations are heading towards two different directions. In a MYC-overexpressing human B-cell line, MYC-mediated cellular proliferation is inhibited by inhibition of E2F1 that is down-regulated by miR-17-92 cluster [49]. Contrary to that tumor-suppressor activity, the same cluster acts as a potential oncogene in B-cell lymphomas by MYC interaction and disabling apoptosis [50]. Therefore, one second theory focuses more on the fact that the same miRNA can take part in distinct pathways depending on gene expression pattern and cell type, hence having not necessarily same effects on cell survival, growth, and proliferation [47]. Giving definitive answers is difficult, because the active molecules that are target of concerning miRNAs can hardly be identified while the state of precursor miRNA expression can. RNA-binding proteins that are abnormally expressed in all cancer types are suspected to be potential interaction partners of miRNA [47, 54].

6 Alterations in the miRNA Processing Machinery

Another theory concerns aberrancies in miRNA-processing genes and proteins. miRNA expression should strongly correlate with alterations of involved proteins. In a study on lung cancer Karube et al. showed that expression levels of Dicer but not Drosha were reduced, leading to a poor chance of survival [55]. Similar reduced survival rates in relation to low Dicer amounts could be identified for let-7 miRNA [56]. Once again, these findings suggest that abnormalities in mature miRNA levels in cancer tissue emerge from aberrant expression as well as aberrant miRNA maturation process. Another study examined expression profiles of Drosha and

Dicer as the two important enzymes in maturation machinery [57]. All types of epithelial tumors presented dysregulated expression levels of either Drosha or Dicer or both of them compared to healthy tissue. Especially for Dicer, many more associations to disease, particularly cancer, have been reported [58]. Interestingly in this context, the transcription factor Sox4, which is associated with certain cancers like melanoma, has been shown to drive Dicer expression [59]. However, up to now it remains quite unclear how such alterations in Dicer expression—usually probed by qRT-PCR of its mRNA—in detail affect human disease. The concentration of miRNA dramatically differs between individual miRNAs and it is unclear if miRNAs expressed at very low concentrations have significant impact on gene regulation. The large differences in expression levels of individual miRNA also suggest that miRNAs expressed at high levels may be more affected by damped dicing activity. In Sox4 knockdown cells, 34.9 % of miRNAs were down-regulated, whereas 26.5 % of miRNAs were up-regulated [59]. A rescue of Dicer expression increased less than two-thirds of the previously down-regulated miRNAs. In fact, it remains to be consistently shown that dicing is a rate-limiting step for the majority of miRNA concentrations. However, a good hint is provided by a study in which siRNA activity, as well as miRNA maturation, was elevated by enoxacin [60]. Enoxacin is a small molecule and belongs to a family of synthetic antibacterial compounds. By using a chemical screen, it was identified to promote RNA interference pathway by enhanced siRNA-mediated mRNA degradation. Furthermore, cells stably expressing a pri-miRNA transcript (pri-miR-125a) showed increase in the mature form of miR-125a after addition of enoxacin. Also, decreases in total levels of pre- and pri-miRNA were observed. Results indicate that enoxacin targets TRBP, which, together with Dicer, processes and loads miRNA onto RISC. Notably, miRNA levels were only elevated when precursor miRNA is generally abundant in the untreated cells. The findings suggest that in those cases the maturation pathway and not transcription of primary miRNA is the rate-limiting step in miRNA biogenesis. These studies underline the theory that maturation machinery is a tightly controlled and by that very slow process that can be enhanced by certain effectors. In addition, enoxacin provides a good example of how small molecules can be used to gradually understand the RNA maturation pathway.

In the Dicer-independent steps of miRNA maturation, it is much more evident how dysregulation can affect only a limited subpopulation of cellular miRNAs. The Drosha-mediated cleavage of pri-miRNA can take place by alternative mechanisms, as this enzyme is a component in multiple complexes that often also contain helicases like p68 and p72 which are implicated in diverse miRNA maturation pathways [61]. Furthermore, selectivity for certain miRNAs or miRNA families is often defined by RNA-binding proteins like the abovementioned lin-28 binding to the pri-miRNA or pre-miRNA hairpin loops [61]. While their direct association in human disease is evident, the profiling of components of the miRNA maturation pathway seems to be generally less specific and predictive than the direct profiling of miRNAs.

7 Therapeutic Approaches

There are different strategies for manipulating miRNA activity with potential therapeutic significance (*see* Chapter 3 and ref. 17). Among the different strategies, so far only the substitution or the mimicking of aberrantly expressed miRNAs as well as the direct complementary binding of mature miRNAs by anti-miRs or antagomirs found their way into clinical trials. Antagomirs are a special variation of antimirs, namely, chemically modified single-stranded RNA analogues targeting mature miRNA [62]. Intravenous administration of mir-122 antagomirs in mice decreased corresponding miRNA levels in several organs in a specific, efficient, and long-lasting manner. 2′-O-methyl oligoribonucleotide antagomirs complementary to mir-122 were first reported to reduce cholesterol levels by down-regulating genes involved into cholesterol biosynthesis which points towards new strategies for gene regulation via miRNA silencing in disease.

The most promising study turned into clinical trial and has finished phase two: the first miRNA-targeting drug for treatment of HCV infection [63]. The drug, called miravirsen, is the anti-miR-122 oligonucleotide [64]. MiR-122 is involved in physiological processes in hepatic function and also in liver pathology. It appears to stimulate HCV replication through unusual interaction with two binding sites of 5′-UTR in HCV genome; thus miR-122 silencing leads to inhibition of replication. On the other hand, loss of miR-122 expression in hepatocellular carcinoma (HCC) points to a poor prognosis and metastasis [65]. Remarkably, in case of chronic lymphocytic leukemia (CLL) several targets of miR-122 could be identified and proven to be involved in hepatocarcinogenesis, among them serum response factor (SRF), insulin-like growth factor 1 receptor, and cyclin G1. This bivalent behavior of miR-122 demonstrates the complexity by which one miRNA can be integrated into cellular pathways. Thereby it becomes an even more interesting target for treatment of CLL and HCC, but might also explain why it proves difficult to achieve effective conventional therapy. There is a lot of effort directed towards miRNA-based cancer treatment by several companies developing therapeutics against different miRNA targets [66].

The search for RNA-based drugs includes not only miRNAs but also antisense oligos, siRNAs, short hairpin RNAs, RNA decoys, ribozymes, splice site-targeted oligos, and aptamers [67].

Most demanding challenge in drug development is to address the desired tissue, because drug delivery is not always efficient enough. Normally, oligos do not cross the blood–brain barrier and do not enter heart or muscle. Several tissues though can be reached by simply injecting the drug into the bloodstream. Another way is the subcutaneous administration, which localizes the drug for example in liver, kidney, and fat. Besides delivery issues, most concerns direct towards off-target effects and immune responses caused by treatment but do not seem to be insuperable. A very elegant possibility is local drug delivery, for example to the eye or the lung. The first RNA-based drug, called fomivirsen, is used to treat inflammation of the eye that is caused by cytomegalovirus. It is an antisense oligo that inhibits target gene expression [68, 69].

Despite using oligonucleotides or sponges targeting *mature* miRNA, some other approaches to interfere with miRNA maturation *itself* are obviating access amounts of miRNA. Hence, upstream regulation may be the even better choice of contact point. One possibility to inhibit Dicer-mediated cleavage by specifically binding pre-miRNAs is realized by finding RNA-binding small molecules [70–72]. Recently, Maiti et al. found a specific inhibitory activity on miR-21 maturation for the antituberculosis drug streptomycin [73]. A completely different, but maybe even more promising, strategy has been developed by Deiters et al. They made use of a luciferase-based miRNA reporter plasmid that after transcription yields an mRNA comprising miRNA-binding site in 3′-UTR [74]. Two independent screens identified small molecules which lead to reduced miRNA expression in living cells and presumably act upstream pri-miRNA formation [75, 76]. Such approaches imply that the drawbacks of nucleic acid-based drugs may be circumvented in the future.

8 Conclusion

To sum up, miRNA maturation is a potent target in cancer diagnostics and therapy, because changes in the expression pattern of miRNA correlate with abnormal cell growing which often leads to cancer. But there also is a variety of other human diseases certain miRNAs are linked to, including viral infections, neurodegenerative diseases, and diabetes. miRNAs involved in cancerogenesis either act as tumor suppressors or apoptotic inhibitors; hence they are aberrant or excrescent in cancer cells, respectively. The origin of abnormal mRNA processing may be found in two different levels of deregulation, which are miRNA transcription and miRNA maturation. Naturally, there is a need for adequate assays to detect changes in miRNA maturation and following change in expression patterns. One step to fully understand correlation between disease and miRNA levels is to determine the abnormal point of change on the way from miRNA expression over miRNA maturation to miRNA targets. Determination of miRNA levels may become *the* indispensable diagnostic tool in cancer recognition. Therapeutic approaches including miRNA targeting are to be understood in terms of how they work to succeed in the clinic. But past efforts that resulted in drug approval demonstrate that it is worth investigating. Not only that RNA-based therapy has its advantage in recognizing one specific type of cancer, which makes personalized therapy possible, but it can also support conventional cancer therapeutics by pre-weakening the tumor. Both diagnostic and therapeutic approaches will benefit from establishing a broad range of assays and protocols to investigate miRNA maturation and its whole pathway closely.

References

1. Almeida MI, Reis RM, Calin GA (2011) MicroRNA history: discovery, recent applications, and next frontiers. Mutat Res. doi:10.1016/j.mrfmmm.2011.03.009
2. Stefani G, Slack FJ (2008) Small non-coding RNAs in animal development. Nat Rev Mol Cell Biol 9(3):219–230. doi:10.1038/nrm2347
3. Lee RC, Feinbaum RL, Ambros V (1993) The C. elegans heterochronic gene lin-4 encodes small RNAs with antisense complementarity to lin-14. Cell 75(5):843–854
4. Ruvkun G, Ambros V, Coulson A, Waterston R, Sulston J, Horvitz HR (1989) Molecular genetics of the Caenorhabditis elegans heterochronic gene lin-14. Genetics 121(3):501–516
5. Wightman B, Burglin TR, Gatto J, Arasu P, Ruvkun G (1991) Negative regulatory sequences in the lin-14 3′-untranslated region are necessary to generate a temporal switch during Caenorhabditis elegans development. Genes Dev 5(10):1813–1824
6. Olsen PH, Ambros V (1999) The lin-4 regulatory RNA controls developmental timing in Caenorhabditis elegans by blocking LIN-14 protein synthesis after the initiation of translation. Dev Biol 216(2):671–680. doi:10.1006/dbio.1999.9523
7. Kato M, Slack FJ (2012) Ageing and the small, non-coding RNA world. Ageing Res Rev. doi:10.1016/j.arr.2012.03.012
8. Reinhart BJ, Slack FJ, Basson M, Pasquinelli AE, Bettinger JC, Rougvie AE, Horvitz HR, Ruvkun G (2000) The 21-nucleotide let-7 RNA regulates developmental timing in Caenorhabditis elegans. Nature 403(6772):901–906
9. Pasquinelli AE, Reinhart BJ, Slack F, Martindale MQ, Kuroda MI, Maller B, Hayward DC, Ball EE, Degnan B, Muller P, Spring J, Srinivasan A, Fishman M, Finnerty J, Corbo J, Levine M, Leahy P, Davidson E, Ruvkun G (2000) Conservation of the sequence and temporal expression of let-7 heterochronic regulatory RNA. Nature 408(6808):86–89. doi:10.1038/35040556
10. Bartel DP (2004) MicroRNAs: genomics, biogenesis, mechanism, and function. Cell 116(2):281–297
11. Siomi H, Siomi MC (2010) Posttranscriptional regulation of microRNA biogenesis in animals. Mol Cell 38(3):323–332. doi:10.1016/j.molcel.2010.03.013
12. Lee Y, Ahn C, Han J, Choi H, Kim J, Yim J, Lee J, Provost P, Radmark O, Kim S, Kim VN (2003) The nuclear RNase III Drosha initiates microRNA processing. Nature 425(6956):415–419. doi:10.1038/nature01957
13. He L, Hannon GJ (2004) MicroRNAs: small RNAs with a big role in gene regulation. Nat Rev Genet 5(7):522–531. doi:10.1038/nrg1379
14. Tomari Y, Zamore PD (2005) Perspective: machines for RNAi. Genes Dev 19(5):517–529. doi:10.1101/gad.1284105
15. Hannon GJ (2002) RNA interference. Nature 418(6894):244–251. doi:10.1038/418244a
16. Arenz C (2006) MicroRNAs—future drug targets? Angew Chem Int Ed Engl 45(31):5048–5050
17. Deiters A (2010) Small molecule modifiers of the microRNA and RNA interference pathway. AAPS J 12(1):51–60. doi:10.1208/s12248-009-9159-3
18. Lu J, Getz G, Miska EA, Alvarez-Saavedra E, Lamb J, Peck D, Sweet-Cordero A, Ebert BL, Mak RH, Ferrando AA, Downing JR, Jacks T, Horvitz HR, Golub TR (2005) MicroRNA expression profiles classify human cancers.

Nature 435(7043):834–838. doi:10.1038/nature03702

19. Pfeffer S, Zavolan M, Grasser FA, Chien M, Russo JJ, Ju J, John B, Enright AJ, Marks D, Sander C, Tuschl T (2004) Identification of virus-encoded microRNAs. Science 304(5671):734–736. doi:10.1126/science.1096781
20. Cullen BR (2013) MicroRNAs as mediators of viral evasion of the immune system. Nat Immunol 14(3):205–210. doi:10.1038/ni.2537
21. Gottwein E, Mukherjee N, Sachse C, Frenzel C, Majoros WH, Chi JT, Braich R, Manoharan M, Soutschek J, Ohler U, Cullen BR (2007) A viral microRNA functions as an orthologue of cellular miR-155. Nature 450(7172):1096–1099. doi:10.1038/nature05992
22. Zhao Y, Xu H, Yao Y, Smith LP, Kgosana L, Green J, Petherbridge L, Baigent SJ, Nair V (2011) Critical role of the virus-encoded microRNA-155 ortholog in the induction of Marek's disease lymphomas. PLoS Pathog 7(2):e1001305. doi:10.1371/journal.ppat.1001305
23. Jopling CL, Yi M, Lancaster AM, Lemon SM, Sarnow P (2005) Modulation of hepatitis C virus RNA abundance by a liver-specific MicroRNA. Science 309(5740):1577–1581. doi:10.1126/science.1113329
24. Calin GA, Ferracin M, Cimmino A, Di Leva G, Shimizu M, Wojcik SE, Iorio MV, Visone R, Sever NI, Fabbri M, Iuliano R, Palumbo T, Pichiorri F, Roldo C, Garzon R, Sevignani C, Rassenti L, Alder H, Volinia S, Liu CG, Kipps TJ, Negrini M, Croce CM (2005) A MicroRNA signature associated with prognosis and progression in chronic lymphocytic leukemia. N Engl J Med 353(17):1793–1801. doi:10.1056/NEJMoa050995
25. Yanaihara N, Caplen N, Bowman E, Seike M, Kumamoto K, Yi M, Stephens RM, Okamoto A, Yokota J, Tanaka T, Calin GA, Liu CG, Croce CM, Harris CC (2006) Unique microRNA molecular profiles in lung cancer diagnosis and prognosis. Cancer Cell 9(3):189–198. doi:10.1016/j.ccr.2006.01.025
26. Iorio MV, Ferracin M, Liu CG, Veronese A, Spizzo R, Sabbioni S, Magri E, Pedriali M, Fabbri M, Campiglio M, Menard S, Palazzo JP, Rosenberg A, Musiani P, Volinia S, Nenci I, Calin GA, Querzoli P, Negrini M, Croce CM (2005) MicroRNA gene expression deregulation in human breast cancer. Cancer Res 65(16):7065–7070. doi:10.1158/0008-5472.CAN-05-1783
27. Mendell JT, Olson EN (2012) MicroRNAs in stress signaling and human disease. Cell 148(6):1172–1187. doi:10.1016/j.cell.2012.02.005
28. Lu M, Zhang Q, Deng M, Miao J, Guo Y, Gao W, Cui Q (2008) An analysis of human microRNA and disease associations. PLoS One 3(10):e3420. doi:10.1371/journal.pone.0003420
29. Zandberga E, Kozirovskis V, Abols A, Andrejeva D, Purkalne G, Line A (2013) Cell-free microRNAs as diagnostic, prognostic, and predictive biomarkers for lung cancer. Genes Chromosomes Cancer 52(4):356–369. doi:10.1002/gcc.22032
30. Mitchell PS, Parkin RK, Kroh EM, Fritz BR, Wyman SK, Pogosova-Agadjanyan EL, Peterson A, Noteboom J, O'Briant KC, Allen A, Lin DW, Urban N, Drescher CW, Knudsen BS, Stirewalt DL, Gentleman R, Vessella RL, Nelson PS, Martin DB, Tewari M (2008) Circulating microRNAs as stable blood-based markers for cancer detection. Proc Natl Acad Sci U S A 105(30):10513–10518. doi:10.1073/pnas.0804549105
31. Chen X, Ba Y, Ma L, Cai X, Yin Y, Wang K, Guo J, Zhang Y, Chen J, Guo X, Li Q, Li X, Wang W, Wang J, Jiang X, Xiang Y, Xu C, Zheng P, Zhang J, Li R, Zhang H, Shang X, Gong T, Ning G, Zen K, Zhang CY (2008) Characterization of microRNAs in serum: a novel class of biomarkers for diagnosis of cancer and other diseases. Cell Res 18(10):997–1006. doi:10.1038/cr.2008.282
32. Valadi H, Ekstrom K, Bossios A, Sjostrand M, Lee JJ, Lotvall JO (2007) Exosome-mediated transfer of mRNAs and microRNAs is a novel mechanism of genetic exchange between cells. Nat Cell Biol 9(6):654–659. doi:10.1038/ncb1596
33. Zernecke A, Bidzhekov K, Noels H, Shagdarsuren E, Gan L, Denecke B, Hristov M, Koppel T, Jahantigh MN, Lutgens E, Wang S, Olson EN, Schober A, Weber C (2009) Delivery of microRNA-126 by apoptotic bodies induces CXCL12-dependent vascular protection. Science Signal 2(100):ra81. doi:10.1126/scisignal.2000610
34. Mittelbrunn M, Gutierrez-Vazquez C, Villarroya-Beltri C, Gonzalez S, Sanchez-Cabo F, Gonzalez MA, Bernad A, Sanchez-Madrid F (2011) Unidirectional transfer of microRNA-loaded exosomes from T cells to antigen-presenting cells. Nat Commun 2:282. doi:10.1038/ncomms1285
35. Vickers KC, Palmisano BT, Shoucri BM, Shamburek RD, Remaley AT (2011) MicroRNAs are transported in plasma and delivered to recipient cells by high-density lipoproteins. Nat Cell Biol 13(4):423–433. doi:10.1038/ncb2210
36. Arroyo JD, Chevillet JR, Kroh EM, Ruf IK, Pritchard CC, Gibson DF, Mitchell PS, Bennett CF, Pogosova-Agadjanyan EL, Stirewalt DL, Tait JF, Tewari M (2011) Argonaute2

complexes carry a population of circulating microRNAs independent of vesicles in human plasma. Proc Natl Acad Sci U S A 108(12): 5003–5008. doi:10.1073/pnas.1019055108
37. Seggerson K, Tang L, Moss EG (2002) Two genetic circuits repress the Caenorhabditis elegans heterochronic gene lin-28 after translation initiation. Dev Biol 243(2):215–225. doi:10.1006/dbio.2001.0563
38. Calin GA, Croce CM (2006) MicroRNA-cancer connection: the beginning of a new tale. Cancer Res 66(15):7390–7394. doi:10.1158/0008-5472.CAN-06-0800
39. Liu CG, Calin GA, Volinia S, Croce CM (2008) MicroRNA expression profiling using microarrays. Nat Protocol 3(4):563–578. doi:10.1038/nprot.2008.14
40. Chan JA, Krichevsky AM, Kosik KS (2005) MicroRNA-21 is an antiapoptotic factor in human glioblastoma cells. Cancer Res 65(14):6029–6033. doi:10.1158/0008-5472.CAN-05-0137
41. Jiang J, Lee EJ, Schmittgen TD (2006) Increased expression of microRNA-155 in Epstein-Barr virus transformed lymphoblastoid cell lines. Genes Chromosomes Cancer 45(1): 103–106. doi:10.1002/gcc.20264
42. Wark AW, Lee HJ, Corn RM (2008) Multiplexed detection methods for profiling microRNA expression in biological samples. Angew Chem Int Ed Engl 47(4):644–652. doi:10.1002/anie.200702450
43. Meyer SU, Pfaffl MW, Ulbrich SE (2010) Normalization strategies for microRNA profiling experiments: a 'normal' way to a hidden layer of complexity? Biotechnol Lett 32(12): 1777–1788. doi:10.1007/s10529-010-0380-z
44. Peltier HJ, Latham GJ (2008) Normalization of microRNA expression levels in quantitative RT-PCR assays: identification of suitable reference RNA targets in normal and cancerous human solid tissues. RNA 14(5):844–852. doi:10.1261/rna.939908
45. Creighton CJ, Reid JG, Gunaratne PH (2009) Expression profiling of microRNAs by deep sequencing. Brief Bioinform 10(5):490–497. doi:10.1093/bib/bbp019
46. Bissels U, Wild S, Tomiuk S, Holste A, Hafner M, Tuschl T, Bosio A (2009) Absolute quantification of microRNAs by using a universal reference. RNA 15(12):2375–2384. doi:10.1261/rna.1754109
47. Calin GA, Croce CM (2006) MicroRNA signatures in human cancers. Nat Rev Cancer 6(11):857–866. doi:10.1038/nrc1997
48. Cole MD, McMahon SB (1999) The Myc oncoprotein: a critical evaluation of transactivation and target gene regulation. Oncogene 18(19):2916–2924. doi:10.1038/sj.onc.1202748
49. O'Donnell KA, Wentzel EA, Zeller KI, Dang CV, Mendell JT (2005) c-Myc-regulated microRNAs modulate E2F1 expression. Nature 435(7043):839–843
50. He L, Thomson JM, Hemann MT, Hernando-Monge E, Mu D, Goodson S, Powers S, Cordon-Cardo C, Lowe SW, Hannon GJ, Hammond SM (2005) A microRNA polycistron as a potential human oncogene. Nature 435(7043):828–833
51. Calin GA, Sevignani C, Dumitru CD, Hyslop T, Noch E, Yendamuri S, Shimizu M, Rattan S, Bullrich F, Negrini M, Croce CM (2004) Human microRNA genes are frequently located at fragile sites and genomic regions involved in cancers. Proc Natl Acad Sci U S A 101(9):2999–3004. doi:10.1073/pnas.0307323101
52. Volinia S, Calin GA, Liu CG, Ambs S, Cimmino A, Petrocca F, Visone R, Iorio M, Roldo C, Ferracin M, Prueitt RL, Yanaihara N, Lanza G, Scarpa A, Vecchione A, Negrini M, Harris CC, Croce CM (2006) A microRNA expression signature of human solid tumors defines cancer gene targets. Proc Natl Acad Sci U S A 103(7): 2257–2261. doi:10.1073/pnas.0510565103
53. Meng F, Henson R, Lang M, Wehbe H, Maheshwari S, Mendell JT, Jiang J, Schmittgen TD, Patel T (2006) Involvement of human micro-RNA in growth and response to chemotherapy in human cholangiocarcinoma cell lines. Gastroenterology 130(7):2113–2129. doi:10.1053/j.gastro.2006.02.057
54. Carpenter B, MacKay C, Alnabulsi A, MacKay M, Telfer C, Melvin WT, Murray GI (2006) The roles of heterogeneous nuclear ribonucleoproteins in tumour development and progression. Biochim Biophys Acta 1765(2):85–100. doi:10.1016/j.bbcan.2005.10.002
55. Karube Y, Tanaka H, Osada H, Tomida S, Tatematsu Y, Yanagisawa K, Yatabe Y, Takamizawa J, Miyoshi S, Mitsudomi T, Takahashi T (2005) Reduced expression of Dicer associated with poor prognosis in lung cancer patients. Cancer Sci 96(2):111–115. doi:10.1111/j.1349-7006.2005.00015.x
56. Takamizawa J, Konishi H, Yanagisawa K, Tomida S, Osada H, Endoh H, Harano T, Yatabe Y, Nagino M, Nimura Y, Mitsudomi T, Takahashi T (2004) Reduced expression of the let-7 microRNAs in human lung cancers in association with shortened postoperative survival. Cancer Res 64(11):3753–3756. doi:10.1158/0008-5472.CAN-04-0637
57. Sand M, Skrygan M, Georgas D, Arenz C, Gambichler T, Sand D, Altmeyer P, Bechara FG (2011) Expression levels of the microRNA maturing microprocessor complex component DGCR8

and the RNA-induced silencing complex (RISC) components Argonaute-1, Argonaute-2, PACT, TARBP1, and TARBP2 in epithelial skin cancer. Mol Carcinog. doi:10.1002/mc.20861

58. Pellegrino L, Jacob J, Roca-Alonso L, Krell J, Castellano L, Frampton AE (2013) Altered expression of the miRNA processing endoribonuclease Dicer has prognostic significance in human cancers. Expert Rev Anticancer Ther 13(1):21–27. doi:10.1586/era.12.150
59. Jafarnejad SM, Ardekani GS, Ghaffari M, Martinka M, Li G (2012) Sox4-mediated Dicer expression is critical for suppression of melanoma cell invasion. Oncogene. doi:10.1038/onc.2012.239
60. Shan G, Li Y, Zhang J, Li W, Szulwach KE, Duan R, Faghihi MA, Khalil AM, Lu L, Paroo Z, Chan AW, Shi Z, Liu Q, Wahlestedt C, He C, Jin P (2008) A small molecule enhances RNA interference and promotes microRNA processing. Nat Biotechnol 26(8):933–940. doi:10.1038/nbt.1481
61. Newman MA, Hammond SM (2010) Emerging paradigms of regulated microRNA processing. Genes Dev 24(11):1086–1092. doi:10.1101/gad.1919710
62. Krutzfeldt J, Rajewsky N, Braich R, Rajeev KG, Tuschl T, Manoharan M, Stoffel M (2005) Silencing of microRNAs in vivo with 'antagomirs'. Nature 438(7068):685–689
63. Lindow M, Kauppinen S (2012) Discovering the first microRNA-targeted drug. J Cell Biol 199(3):407–412. doi:10.1083/jcb.201208082
64. Hu J, Xu Y, Hao J, Wang S, Li C, Meng S (2012) MiR-122 in hepatic function and liver diseases. Protein Cell 3(5):364–371. doi:10.1007/s13238-012-2036-3
65. Karakatsanis A, Papaconstantinou I, Gazouli M, Lyberopoulou A, Polymeneas G, Voros D (2011) Expression of microRNAs, miR-21, miR-31, miR-122, miR-145, miR-146a, miR-200c, miR-221, miR-222, and miR-223 in patients with hepatocellular carcinoma or intrahepatic cholangiocarcinoma and its prognostic significance. Mol Carcinog. doi:10.1002/mc.21864
66. Nana-Sinkam SP, Croce CM (2013) Clinical applications for microRNAs in cancer. Clin Pharmacol Ther 93(1):98–104. doi:10.1038/clpt.2012.192
67. Bonetta L (2009) RNA-based therapeutics: ready for delivery? Cell 136(4):581–584. doi:10.1016/j.cell.2009.02.010
68. Highleyman L (1998) Fomivirsen. BETA 29: 31
69. Razonable RR (2011) Antiviral drugs for viruses other than human immunodeficiency virus. Mayo Clin Proc 86(10):1009–1026. doi:10.4065/mcp.2011.0309
70. Davies BP, Arenz C (2006) A homogenous assay for micro RNA maturation. Angew Chem Int Ed Engl 45(33):5550–5552
71. Davies BP, Arenz C (2008) A fluorescence probe for assaying micro RNA maturation. Bioorg Med Chem 16(1):49–55
72. Neubacher S, Dojahn CM, Arenz C (2011) A rapid assay for miRNA maturation by using unmodified pre-miRNA. ChemBioChem 12(15):2302–2305. doi:10.1002/cbic.201100445
73. Bose D, Jayaraj G, Suryawanshi H, Agarwala P, Pore SK, Banerjee R, Maiti S (2012) The tuberculosis drug streptomycin as a potential cancer therapeutic: inhibition of miR-21 function by directly targeting its precursor. Angew Chem Int Ed Engl 51(4):1019–1023. doi:10.1002/anie.201106455
74. Connelly CM, Thomas M, Deiters A (2012) High-throughput luciferase reporter assay for small-molecule inhibitors of microRNA function. J Biomol Screen. doi:10.1177/1087057112439606
75. Gumireddy K, Young DD, Xiong X, Hogenesch JB, Huang Q, Deiters A (2008) Small-molecule inhibitors of microrna miR-21 function. Angew Chem Int Ed Engl 47(39):7482–7484. doi:10.1002/anie.200801555
76. Young DD, Connelly CM, Grohmann C, Deiters A (2010) Small molecule modifiers of microRNA miR-122 function for the treatment of hepatitis C virus infection and hepatocellular carcinoma. J Am Chem Soc 132(23):7976–7981. doi:10.1021/ja910275u

Chapter 3

Approaches to the Modulation of miRNA Maturation

Valerie T. Tripp, Jaclyn R. McKenna, and Douglas D. Young

Abstract

The therapeutic importance of microRNA (miRNA) regulation has recently been realized as these small, noncoding RNAs have been demonstrated to be involved with a plethora of diseases and disorders. Due to the complex miRNA maturation process, the expression of these important biomolecules can be manipulated at various stages of the pathway. This review examines both in vivo and in vitro mechanisms and assays that have been developed to regulate miRNA levels. Modulation of miRNA maturation can be accomplished via several therapeutic agents, including small molecules and oligonucleotides, in both specific and nonspecific fashions. Due to the relevance of miRNAs, these novel therapeutic approaches represent new tools for the treatment of various cancers and other deleterious disorders.

Key words miRNA maturation, Small-molecule therapeutics, miRNA assays, Antisense agents

1 Introduction

Knowledge about the biological relevance of microRNAs (miRNAs) has grown exponentially in the past several years, specifically their involvement in gene regulation and disease progression [1–5]. These small 21–23mer RNA nucleotides were long thought to be cellular "junk"; however, they have now been demonstrated to post-transcriptionally modulate gene expression by site-specifically binding mRNA transcript [6–10]. Over 900 human miRNA genes have been reported, and it is thought that they regulate over one-third of all human genes [11, 12]. Their importance has been demonstrated in cell proliferation, cell differentiation, and various developmental processes as well as a range of disease states including various forms of cancer and viral replication cycles [13–18]. Typically, misregulation of miRNA levels can be correlated to several cancers and other disorders [19–22]. The restoration (typically by antisense technologies) of normal miRNA levels has been demonstrated to correct the disease state, often by induction of apoptosis of the diseased cells [23–25]. Consequently, miRNAs

Christoph Arenz (ed.), *miRNA Maturation: Methods and Protocols*, Methods in Molecular Biology, vol. 1095,
DOI 10.1007/978-1-62703-703-7_3, © Springer Science+Business Media New York 2014

have become extremely important non-conventional drug targets for therapeutic intervention [4, 5, 26–29].

The biogenesis of miRNAs involves nuclear transcription by RNA polymerase II of endogenously encoded primary miRNAs (pri-miRNAs) 100–1,000 nucleotides in length (Fig. 1) [10, 30]. Pri-miRNAs are then further processed in the nucleus into stem–loop-structured precursor-miRNAs (pre-miRNAs) of 60–80 nucleotides in length, which are exported into the cytoplasm [31, 32]. At this stage pre-miRNAs behave in an analogous fashion to exogenously introduced double-stranded RNAs used to produce siRNAs as pre-miRNAs are further digested to 21–23 nucleotide mature miRNAs by the enzyme Dicer (in conjunction with partnering proteins) and are loaded into the RNA-induced silencing complex (RISC) [33, 34]. At this stage the RNA is unwound and the active miRNA strand is used to site-specifically repress mRNA translation. While the exact mechanism of translational repression has yet to be fully elucidated

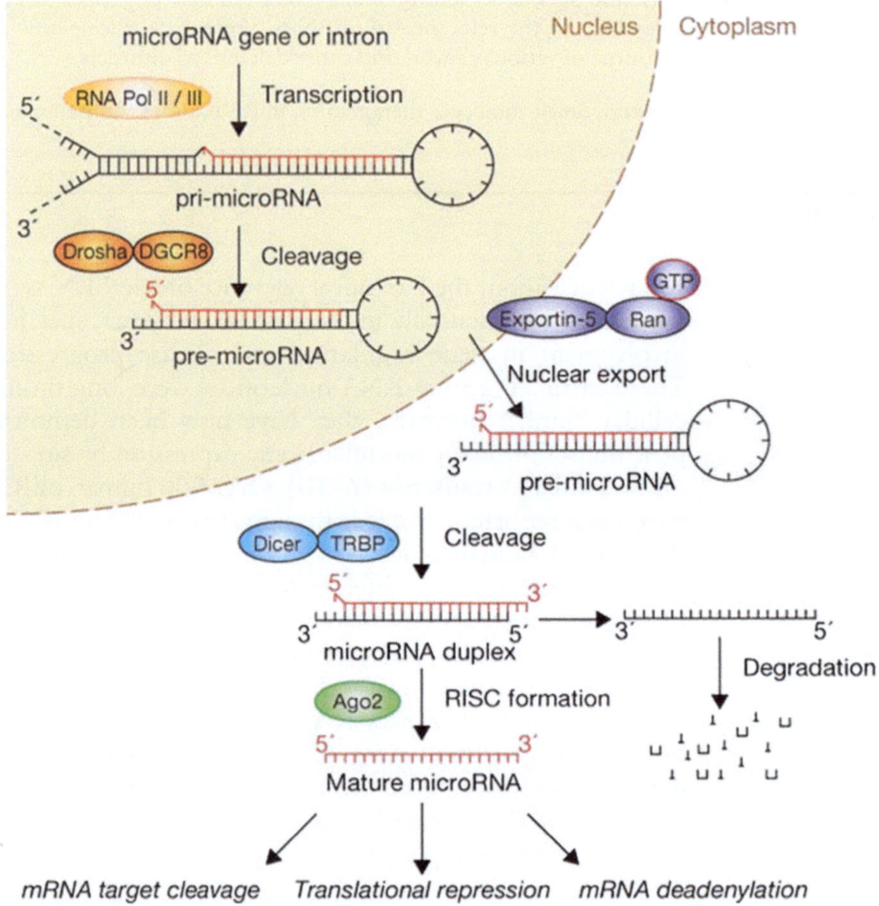

Fig. 1 The representative miRNA maturation pathway leading to gene silencing. Image adapted with permission from Winter J et al. (2009) Nat Cell Biol

several hypotheses exist including mRNA cleavage, deadenylation, degradation, p-body recruitment, and blocking of various translational initiation steps [11, 35].

Due to the complex nature of this biogenesis cascade, there are several sites for therapeutic intervention; however this chapter specifically focuses on the regulation of miRNA maturation rather than simply targeting the mature miRNA [36]. Depending on the target and type of therapeutic this can be achieved in either a specific fashion (only affecting a single miRNA) or a nonspecific fashion (affecting the entire pathway and thus all cellular miRNAs). Both approaches have distinct advantages and disadvantages, as specific targeting is often challenging to achieve but has minimized deleterious off-target effects [37]. Conversely, global targeting is easier to accomplish but requires additional considerations to deliver the therapeutic to only the desired cells in order to avoid miRNA misregulation in healthy cells. In either case, the ability to target miRNA maturation represents not only a viable approach to achieving a better understanding of the miRNA pathway but also a useful mechanism to treat conditions associated with misregulated miRNA and thus misregulated gene expression.

Moreover, the elucidation of valid maturation inhibitors can be accomplished via the development of either in vitro or in vivo assays/screens. While in vitro assays are often easier to establish and more directly target a specific component of the maturation pathway, they can sometimes be difficult to recapitulate when the active hits are translated to an in vivo system [38]. On the other hand, in vivo assays are often more complex to establish and require a deconvolution step to ascertain the actual target of the effector molecule; however, they are already established within biological systems and deliver more robust "hits." [39] This chapter examines both in vitro and in vivo approaches to miRNA maturation at each step in the biogenesis process to highlight the innovative research that has been conducted, affording new chemical tools to control this important biological process.

2 Manipulation of Pri-miRNA Maturation

The ability to modulate the level of pri-miRNA either by transcriptional regulation or direct binding to the tertiary structure of pri-miRNA represents a mechanism to achieve a higher level of specificity than targeting further downstream in the biogenesis pathway. Consequently, targeting of pri-miRNA can be achieved via both in vivo and in vitro assays depending on the choice of regulatory molecule (small molecule or biomacromolecule). Preventing the further maturation of pri-miRNA affords the ability to significantly decrease cellular concentrations of miRNAs and thus has significant therapeutic value.

2.1 Small-Molecule Regulators In Vivo

Regulation of miRNA levels via small-molecule modulators has recently been demonstrated. The utilization of small-molecule therapeutics has several significant advantages over oligonucleotides, most notably substantially improved pharmacokinetic properties [27]. Additional advantages typically include more facile synthetic production and easier cellular delivery/targeting [28]. Moreover, the ability to screen large compound libraries for their ability to affect the entire miRNA biogenesis and regulatory pathway represents a unique mechanism to potentially increase the degree of target specificity that has previously hindered other small molecule approaches.

In order to screen small-molecule libraries for activity, a robust assay is required that facilitates screening of cells in a high-throughput fashion. Work in the Deiters and Huang labs has made substantial progress towards this requirement [40]. Initial screening attempted to ascertain an inhibitor of miR-21 due to its overexpression in a variety of cancers, including glioblastomas. The original assay was developed using a stably transfected luciferase reporter system with a miRNA-binding site cloned into the 3′ UTR of the luciferase gene. Various constructs (including a miR-21-, miR-30-, or non-miRNA-binding site) were stably introduced into HeLa cells due to their well-reported misregulation of miR-21. In the absence of small molecules, cells harboring the miR-21-binding site construct exhibited a significant decrease in luciferase expression relative to the other constructs due to the overexpression of miR-21 knocking down the luciferase expression. A similar result could be achieved in the miR-30 line via the exogenous introduction of miR-30; however, due to its minimal levels in HeLa cells a significant effect on luciferase levels was not observed without this addition. Due to the ability of the assay to knock down luciferase expression, inhibition of the miRNA pathway at any point should lead to a notable decrease in the miRNA levels and consequently increase the levels of luciferase, providing a positive readout for the assay (Fig. 2). Thus, when cells were grown in the presence of a small-molecule inhibitor, compounds could be easily identified as miRNA effectors. Screening ~1,000 compounds resulted in the discovery of **1** which increased luciferase expression approximately 500 % relative to a DMSO-treated control (Fig. 3). In order to identify the mechanism of action, the miRNA and pri-miRNA levels were examined by qRT PCR, demonstrating decreased miR-21 and pri-miR-21 levels. Interestingly, **1** had no effect on the levels of a variety of other miRNAs, indicating a degree of miRNA specificity. Unfortunately, this compound did not result in the apoptosis of a glioblastoma cell line, and its actual mechanism of action has yet to be established.

While the previously described assay was useful for the identification of small-molecule miRNA inhibitors, several issues precluded its true utilization in a high-throughput screen, including

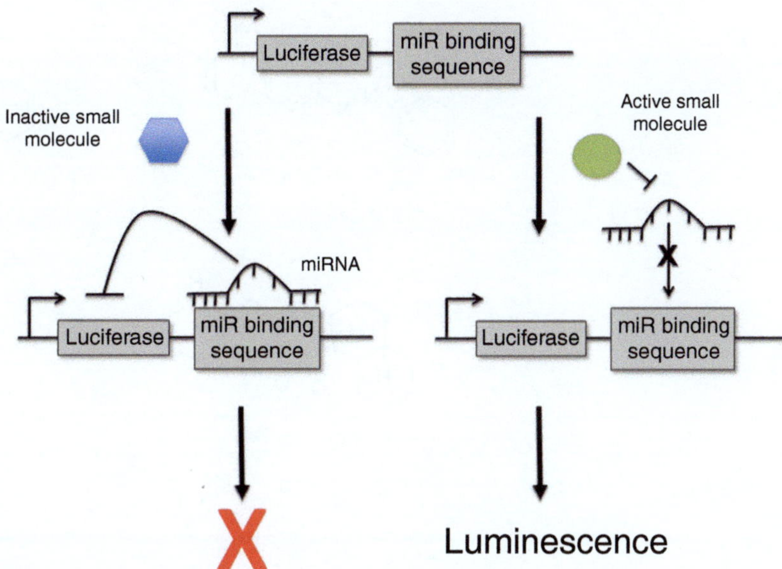

Fig. 2 Cell-based assay for small-molecule miRNA modulators. Using either a GFP or a luciferase reporter with a miRNA-binding sequence in the 3′ UTR. Inactive molecules fail to inhibit miRNA binding and suppression of the gene, resulting in a lack of signal from the reporter protein. Active molecules inhibit miRNA maturation or function abrogating binding, resulting in increased levels of reporter protein expression

the establishment of stable cell lines and the lack of a cellular viability control. In order to rectify these issues, Deiters et al. modified the assay to employ a psiCHECK-2 plasmid that was transiently transfected into the cell line of choice [41]. Moreover, the plasmid contains two luciferase genes, with the miRNA-binding site only in the 3′ UTR of the *Renilla* luciferase, leaving the firefly luciferase to act as both a transfection and cellular viability control. To investigate this assay system, miR-122 was selected due to its overexpression in hepatitis C infections and under-expression in hepatocellular carcinomas. The improved assay yielded two miR-122 inhibitors, **2** and **3**, determined by a reproducible increase in luciferase expression. Moreover, the assay also afforded a miR-122 activator (**4**) as the control reporter system allowed for an effective measurement of decreased levels of luciferase expression (Fig. 3). Both the inhibition and activation of the compounds were confirmed by qRT PCR assays, indicating alterations in levels of both the miR-122 and pri-miRNA-122. Again, the assay indicated some specificity for miR-122, suggesting some level of transcriptional control as opposed to regulation of the general miRNA biogenesis pathway. Gratifyingly, compounds **2** and **3** were able to reduce viral loads of HCV, while **4** was able to induce apoptosis in a hepatocellular carcinoma cell line, suggesting the potential therapeutic values of these miRNA-regulating compounds. However, additional

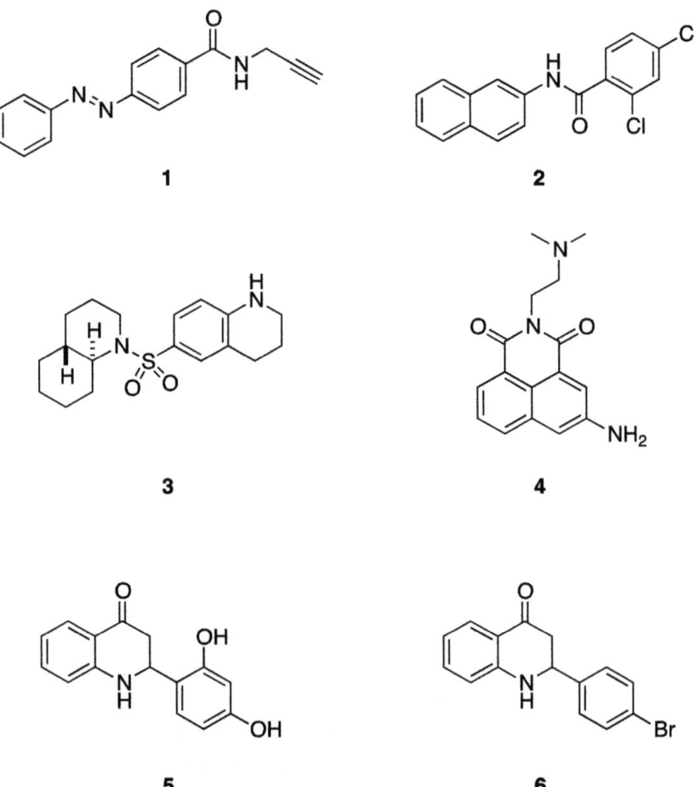

Fig. 3 Small molecules discovered by cell-based screens, which modulate miRNA activity. While the mechanism of action has not been elucidated, many are thought to function via regulation of pri-miRNA transcription

research must still be accomplished to elucidate their mechanism of action and characterize their degree of specificity.

Using a similar dual-luciferase assay, studies in the Bhadra lab identified a group of aza-flavanones capable of inhibiting *Drosophila* miR-14 and its human homolog miR-4644 [42]. After screening nine compounds, **5** and **6** resulted in an ~400 % increase in luciferase signal (Fig. 3), which was confirmed by qRT-PCR assays and cell cytotoxicity assays (albeit only in a twofold differential toxicity relative to healthy cells). The activity of the two compounds was then translated to a *Drosophila* system and used to increase reporter GFP expression in larval wing discs carrying a sensor system which harbored miR-14-binding sites in the 3′ UTR of a GFP gene (Fig. 4). However, the amount of EGFP expression assessed by gel electrophoresis did not appear to be substantially increased by the presence of the compounds.

Interestingly, all three studies afforded compounds that ultimately altered pri-miRNA levels (most likely by transcriptional repression) despite being screened against the entire miRNA biogenesis pathway. Further studies are required to ascertain the

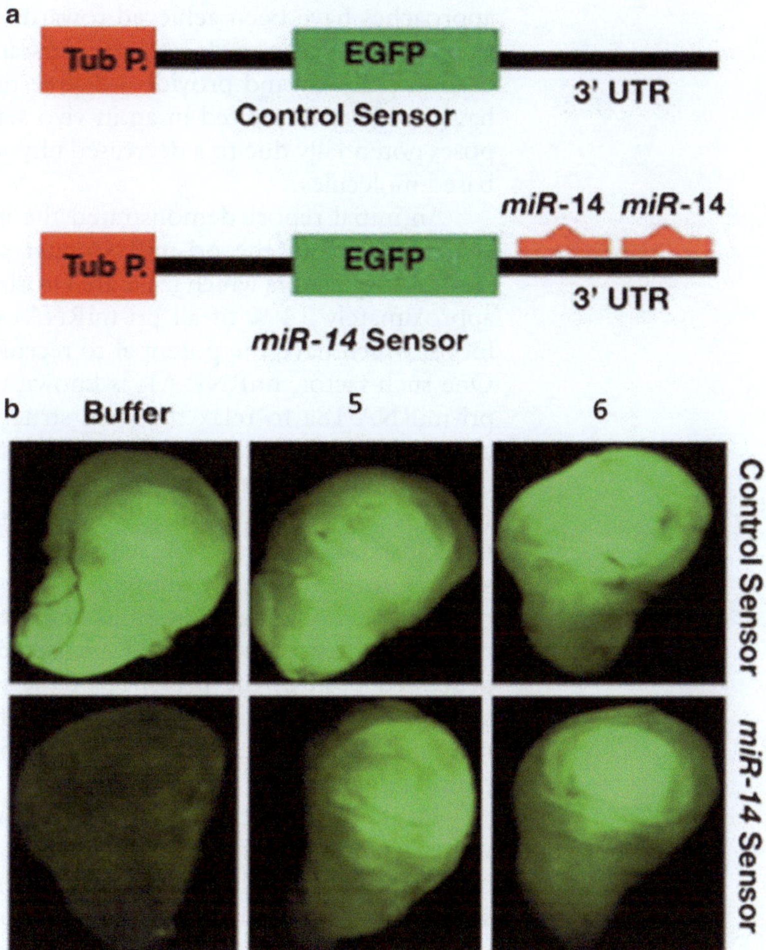

Fig. 4 Aza-flavone inhibition of miR-14 in larval wing discs. (**a**) Schematic of the reporter constructs with either a control sequence with no sequence homology to a miRNA in the 3′ UTR of EGFP or a functional reporter with two miR-14-binding sites in the 3′ UTR. (**b**) EGFP expression in larval discs harboring either the control sensor or the miR-14 sensor in the presence and absence of compounds **5** and **6**. EGFP expression is significantly higher in the miR-14 construct treated with one of the small-molecule inhibitors. Image adapted with permission from Chandrasekhar S et al. (2012) Bioorg Med Chem Lett

true efficacy of these compounds in more relevant in vivo disease models; however, this research effectively demonstrates that lowering the cellular concentrations of pri-miRNAs with small molecules can affect mature miRNA levels and exert a biological consequence, specifically towards cancer cell viability.

2.2 Inhibition of Drosha-Mediated Pri-miRNA Maturation In Vitro

The inhibition of pri-miRNA maturation is also feasible via the disruption of Drosha processing in the nucleus, inhibiting the formation of pre-miRNA. While no small-molecule inhibitor of this processing has been discovered, different oligonucleotide approaches have been achieved towards this goal. While both are extremely useful towards better understanding of the miRNA maturation pathway and provide a degree of miRNA specificity, they have not been exploited in an in vivo setting for therapeutic purposes potentially due to a decreased physiological stability of RNA-based molecules.

An initial report demonstrated the importance of apical-loop structures within the pri-miRNA that serve as binding sites for *trans*-acting factors which facilitate Drosha processing [43]. In fact, approximately 14 % of all pri-miRNAs possess highly conserved loops, which have the potential to recruit these auxiliary proteins. One such factor, hnRNP A1, is known to bind an apical-loop on pri-miRNA-18a to relax the stem structure and facilitate Drosha cleavage [44]. In order to inhibit this binding event, 2′-O-methyl oligonucleotides complementary to this loop (LooptomiRs) were tested to identify the necessity of the binding in miRNA maturation. Interestingly, miRNAs with highly conserved loop domains, such as pri-miR-18, demonstrated a lack of processing to pre-miRNA when incubated in HeLa cell extracts in vitro, while other miRNAs not possessing highly conserved loops were able to be processed even in the presence of the looptomiR. These results suggest that there is a higher level of complexity to Drosha-mediated processing, and the apical loops represent unique therapeutic targets.

The same interaction of hnRNP A1 with pri-miR-18 transcript was also targeted using RNA aptamers as an alternative oligonucleotide strategy to inhibit miRNA biogenesis [45]. Aptamers are simply highly structured RNAs capable of recognizing and specifically binding to a target, and are typically evolved for target recognition via an in vitro selection [46]. Twelve rounds of a selection strategy using an RNA library comprising a 25 nucleotide random region against a 791 nucleotide miRNA polycistron comprising pri-miR-17-18a-19a-20a-19b-1-92 led to several consensus sequence motifs. Surface plasmon resonance (SPR) revealed a 9 nM dissociation constant for the binding of a specific RNA aptamer and the corresponding pri-miRNA cluster. Further studies elucidated a high-affinity binding of the aptamer to the apical-loop domain of pri-miR-18a and a secondary low-affinity binding site between pri-miR-19a and pri-miR20a.

To investigate the interaction with the apical-loop domain, various mutants of miR18a were constructed and observed in the presence of the aptamer. None of the mutants showed measurable binding with the RNA aptamer, further indicating the importance of this region. Additional RNA footprinting experiments revealed that the presence of the evolved aptamer results in protection of the apical-loop domain from Drosha-mediated degradation. In contrast to the previous looptomiR inhibition, the aptamer also inhibited the processing of other pri-miRNAs in the cluster, while the looptomiR only affected pre-miR-18, suggesting a slightly different mechanism of action. Additionally, due to the aptamer being selected against the entire polycistron, some degree of miRNA specificity is lost.

The use of aptamers to interfere with miRNA maturation in vitro shows promise as a tool to modulate miRNA function. Such methods allow for more specific means to elucidate the miRNA processing and maturation pathways. However, in order to be truly effective, the action of these RNA aptamers must also be demonstrated to be functional within in vivo studies involving more complex environments.

3 Manipulation of Pre-miRNA Maturation

The targeting of pre-miRNA maturation has received significantly more attention, potentially due to a greater wealth of information surrounding both pre-miRNA precursors and their processing enzymes (e.g., Dicer) as a result of their similarities to the siRNA maturation pathways. Moreover, the stem–loop structure of pre-miRNAs makes them amenable to the development of both in vivo and in vitro assays using both small molecules and biomacromolecules. The processing of pre-miRNAs to mature miRNAs can also be modulated by a variety of mechanisms including direct binding as well as inhibition of Dicer processing.

3.1 Specific Binding of Pre-miRNA by Peptides and Peptoids In Vitro

The ability to develop specific non-oligonucleotide binders to RNA is a challenging task due to the minimal chemical diversity and secondary structure of small RNAs; however, several researchers have exploited both peptides and peptoids towards this goal. Research in the Beal laboratory investigated a series of helix-threading peptides (HTPs) harboring an intercalating heterocyclic core and peptide appendages capable of stabilizing interactions with a target RNA [47]. There are two primary advantages for employing macrocyclic HTPs to target RNA, firstly that HTPs offer a distinct orientation of peptide functional groups. This structural rigidity provides a recognition surface with limited conformational flexibility. However, although these conformations are limited, not all conformations confer specific amino acid–nucleotide interactions. The second major advantage is the increased cellular permeability, stability, and RNA binding affinity of cyclic peptides relative to their linear analogs.

HTPs were prepared by a relatively short solid-supported synthesis of a protected quinoline acid containing diallyl peptide followed by a ring closing metathesis to generate macrocyclic peptides (Scheme 1). Only two cyclic HTPs were reported, **7** and **8**, with their binding properties assessed against a targeted stem–loop RNA by ribonuclease footprinting. Although one of the resulting

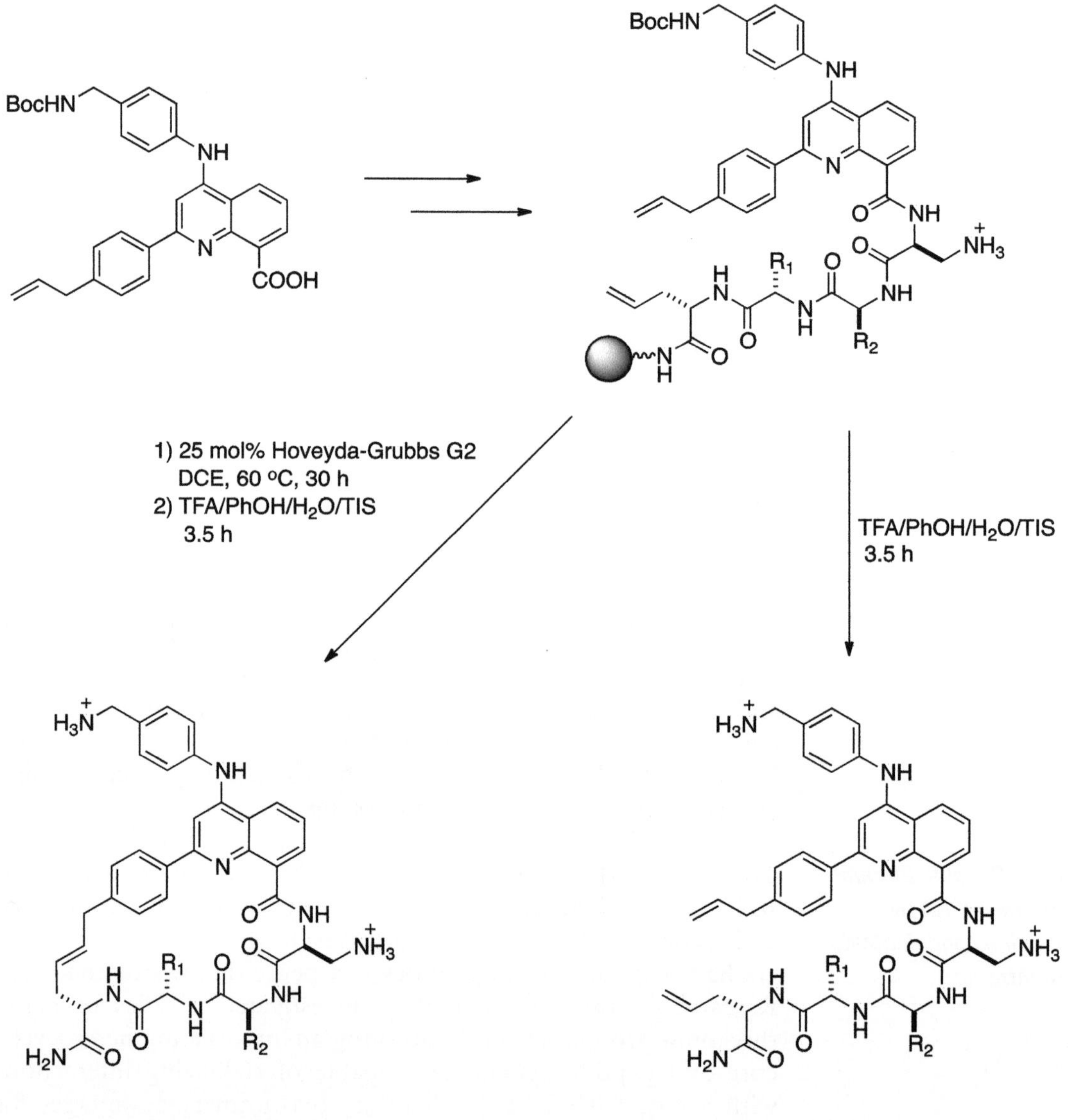

1) 25 mol% Hoveyda-Grubbs G2
DCE, 60 °C, 30 h
2) TFA/PhOH/H_2O/TIS
3.5 h

TFA/PhOH/H_2O/TIS
3.5 h

7 : R_1 = H, R_2 = $(CH_2)_4NH_3^+$
8 : R_1 = $(CH_2)_4NH_3^+$, R_2 = H

9 : R_1 = H, R_2 = $(CH_2)_4NH_3^+$

Scheme 1 Synthesis of macrocyclic helix-threading peptides

HTPs had a much lower dissociation constant (52 and 271 nM), both HTPs bound more effectively than a linear counterpart (**9**, 329 nM). Additionally, the macrocyclic HTP was proven capable in distinguishing between different RNA targets by binding efficiency of approximately two orders of magnitude. HTP affinity for a pre-miRNA23b was then examined due to its 5′-PyPu-3′ intercalation site flanked by 3′ bulges, and found to be 2.3 μM, indicating the applicability of these HTPs to miRNA maturation. Optimally, this binding interaction should function in a comparable fashion to the footprinting experiment, inhibiting Dicer processing of the pre-miRNA. However, actual testing of maturation was not examined, nor was the specificity of the HTP towards other similar pre-miRNAs. The generation of a larger library of macrocyclic HTPs to increase specificity in conjunction with the in vivo assessment of these compounds may lead to viable inhibitors of specific pre-miRNA maturation.

In an alternative approach by Yu and co-workers, amphiphilic helical peptides containing acridines were employed as potential pre-miRNA binders [48]. In order to enhance binding affinity and specificity to different target RNA motifs conjugated ligands containing a variety of binding modes were utilized. An acridine moiety was selected to intercalate, while the peptide targets the RNA stem regions. An array of ten amphiphilic peptides with differing α-helical content and an acridinylated residue at variable positions was measured for a binding to different RNA substrates (including pre-miR-24-1) using fluorescence anisotropy. The position of the acridine moiety was found to significantly affect the conformational rigidity, and resulted in differential binding of RNA targets at nanomolar concentrations. By increasing the number of acridinylated residues while maintaining the helicity, the specificity for different targets was enhanced to ~250 pM. However, the specificity for different analogs of pre-miR-24-1 was not discernable, and no other miRNA targets were examined. While this approach is promising, it is still unclear whether the necessary specificity towards a variety of pre-miRNAs can be achieved, and the ability of helical amphiphilic peptides to actually modulate miRNA maturation has yet to be demonstrated either in vivo or in vitro.

Finally, the Luebke laboratory demonstrated the in vitro specificity of a peptoid library to bind pre-miR-21 using a fluorescence assay [49]. Peptoids are useful targeting agents due to their cell permeability, protease resistance, chemical diversity, and ease of synthesis [50]. Additionally, the ability to screen these molecules in a microarray format affords the ability to examine thousands of compounds simultaneously as replicate arrays can be printed from a single chemical library. This feature allows for the screening of the same library against multiple targets. A spatially isolated peptoid library was constructed on a solid support from a chemically diverse set of 21 monomers theoretically representing 9,261 compounds, all with a C-terminal arginine to enhance RNA binding. The peptoid microarrays were then incubated in the presence of fluorescently labeled pre-miR-21 or a fluorescently labeled control RNA to ascertain differential versus nonspecific binding. Positive binding compounds were identified by fluorescence microscopy based on reproducibly detectable fluorescence in the pre-miR-21 conditions versus a control RNA hairpin (Fig. 5). Two peptoids demonstrated specific binding activity with dissociation constants of approximately 2 μM and 20-fold discrimination between the

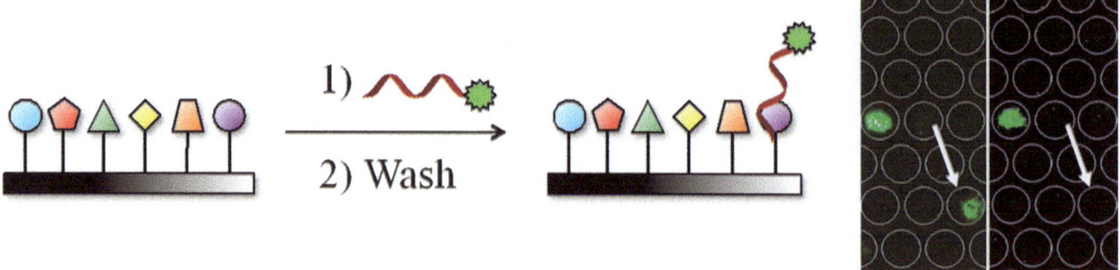

Fig. 5 Functional assay for the screening of immobilized peptoid miRNA binders. Fluorescently labeled miRNAs and RNA controls are incubated with the peptoid library and washed, and peptoids with a high affinity retain the labeled RNA. Binding can be assessed by fluorescence microscopy to detect the labeled RNA. Specificity of the binding can be assessed by comparing plates incubated with the miRNA to the plates incubated with RNA control

different RNA targets. While the strategy readily lends itself to the rapid synthesis of diverse libraries, the requisite for immobilization to a solid-support becomes somewhat limiting. Moreover, while peptoid binding will most likely inhibit pre-miRNA processing, it has yet to be demonstrated in an in vivo or an in vitro context. However, future work probing the specificity of this diverse library against other pre-miRNAs and even pri-miRNAs may yield highly stable compounds that can readily be employed towards the inhibition of miRNA maturation.

3.2 Specific Binding of Pre-miRNA by Peptides and Peptoids In Vivo

Though many research approaches aim to inhibit miRNA function through the use of oligonucleotides or peptides to obtain target specificity in vivo, recent work has strived to employ small molecules to impede miRNA maturation. Such techniques provide several advantages over antagomir and other peptide methods, including more effective delivery to tissues. One such class of small molecules, known as aminoglycosides, is known to bind to secondary structures within RNA, such as stem–loops. Since miRNA precursors contain stem–loops as well as bulges, aminoglycoside-based therapeutics represent potentially potent inhibitors of the miRNA biogenesis pathway.

Using a firefly luciferase reporter system fused to a miRNA-21 target, researchers screened 15 aminoglycosides for their ability to decrease miR-21, a miRNA overexpressed in multiple cancers [51]. Of the molecules, streptomycin acted as a highly effective, selective inhibitor of miR-21 activity at a concentration comparable to inhibition achieved with antisense LNAs against miR-21 (10 μM). Moreover, the presence of streptomycin also resulted in increased levels of PDCD4, a known target of miR-21. Interestingly, a closely related aminoglycoside, dihydrostreptomycin, had no effect on PDCD4 levels in the cells. Thermal melting experiments were used to further explore the mode of action of streptomycin-mediated inhibition. Addition of streptomycin led to thermal stabilization of pre-miR-21, indicating an interaction with the miRNA precursor that may interfere with miRNA maturation (Fig. 6). When screened by RT-PCR against an array of nine miRNAs the streptomycin was found to be relatively specific, with only miR-27a also being downregulated.

Further docking studies between aminoglycosides and pre-miR-21 revealed direct hydrogen bonding between streptose groups and the inner walls of the RNA-binding pocket close to the stem–loop. Given this interaction, researchers chose to investigate the cleavage activity of Dicer after the addition of streptomycin. The presence of streptomycin did selectively decrease the levels of mature miR-21, however only when a specific bulge near the terminal loop was available for streptomycin binding. To probe the effects of streptomycin binding in vivo, a firefly luciferase reporter with the 3′ UTR of PDC4 wild-type pre-miR-21 and mutant pre-miRNA-21 with a deleted bulge were employed. Upon addition of streptomycin, a 1.2-fold increase in luciferase was observed when comparing the wild-type and mutant pre-miRNAs revealing the necessity of the bulge for binding and Dicer processing. Since PDCD4 plays a role in inducing apoptosis, researchers investigated the effects of streptomycin-mediated inhibition of miRNA on cell viability. Streptomycin-treated MCF-7 cancer cells known to over-express miR-21 showed an increase in apoptosis consistent with an increase in PDCD4 levels, while Jurkat cells possessing normal miR-21 levels were unaffected (Fig. 6). Ultimately, streptomycin efficiently represses mature miR-21 levels through direct binding to precursor molecules, thus interfering with Dicer cleavage.

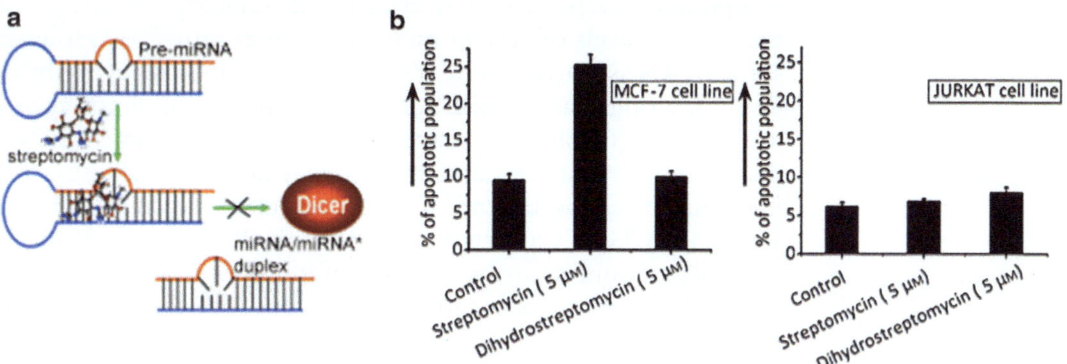

Fig. 6 Streptomycin-modulated inhibition of pre-miRNA maturation. (**a**) Principle of the assay, as streptomycin binding to pre-miRNA inhibits Dicer processing to the mature miRNA, preventing gene silencing. (**b**) FACS analysis of two cell lines treated with streptomycin or dihydrostreptomycin. MCF-7 cells with misregulated miRNA levels exhibit a higher degree of apoptosis when treated with streptomycin, indicating its effect on the miRNA maturation pathway. Healthy JURKAT cells are not affected by streptomycin treatment and thus exhibit normal levels of apoptosis. Image adapted with permission from Bose D et al. (2012) Angew Chem Int Ed Engl

Though streptomycin appears to be specific for miR-21, other small molecules may be found in a larger screen. This exciting example demonstrates that small-molecule binding of pre-miRNAs is indeed biologically relevant and has the potential to inhibit Dicer processing and further miRNA maturation, ultimately modulating the function of the miRNA within the cell. Future studies must be conducted to further optimize the specificity and ascertain if the streptomycin is engaging in off-target interactions within the complex cellular networks.

Another method for inhibiting miRNA maturation through precursor targeting involves the use of morpholino oligonucleotides [52]. These antisense molecules provide a quick method to knock down miRNA during embryonic development and can be easily employed in model organisms, e.g., zebra fish. However, these antisense agents must also be used with some degree of caution as off-target phenotypes can often be observed. Injection of morpholinos complementary to mature miRNAs suppresses miRNA levels for up to 4 days. Using GFP reporters fused to specific miRNA target sites, Plasterk and co-workers found that addition of miRNA silenced the GFP signal, as expected. However, addition of a miRNA duplex and the complementary morpholino restores the GFP reporter, revealing that morpholinos can block the activity of mature miRNA duplexes. Since the addition of morpholino oligonucleotides did not alter miRNA isolation or stability, the interaction appears to inhibit miRNA maturation.

As previously noted, the use of antisense oligonucleotides often results in off-target effects resulting in a different phenotype than the desired knockout. To compare the effects of morpholinos, several morpholinos were designed to target the Drosha and Dicer cleavage sites of precursor miRNAs. These precursor-targeting morpholinos were extremely effective at inducing a reduction in the cellular levels of certain miRNAs, such as miR-205. However some miRNAs, such as miR-30c, were only inhibited by morpholinos targeting mature miR-30c and not the morpholinos against miR-30c precursors. The discrepancy may occur as some miRNAs display more sequence variability and may not be equally susceptible to morpholinos designed to target precursor molecules (Fig. 7). Utilizing a GFP construct with pri-miR-205 in the 3′ UTR, researchers found that the addition of morpholinos resulted in an accumulation of primary miRNA consistent with the inhibition of Drosha activity.

Further work with morpholinos centered on the inhibition of miR-375, a miRNA that has two copies in the zebra fish genome and is expressed in the pancreatic islet as well as the pituitary gland. Multiple morpholinos were designed to target pri-miR-375-1, pri-miR-375-2, or mature miR-375. The morpholinos targeting miRNA precursors were able to interfere with Drosha or Dicer cleavage steps, thus decreasing the levels of mature miR-375. In comparison,

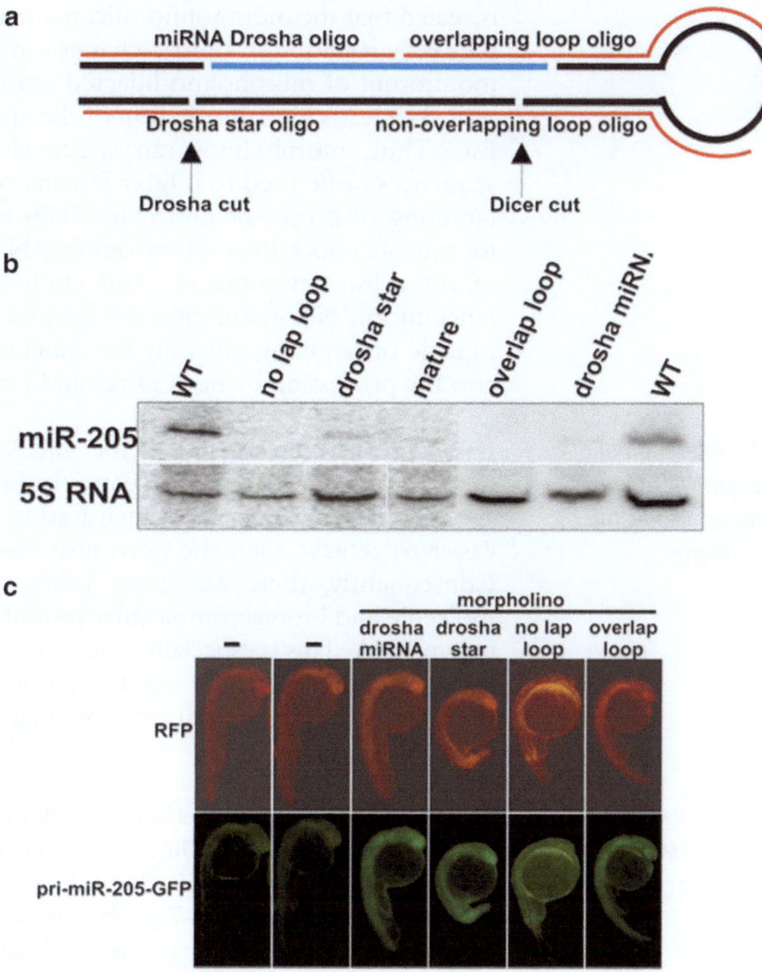

Fig. 7 Morpholino targeting of miRNA maturation in zebra fish. (**a**) Schematic of pri-miRNA with different morpholinos targeting different portions of the sequence, affecting both Drosha and Dicer processing. (**b**) Effect of the different morpholinos on the miR-205 levels assessed by northern blot analysis of 30-h-old embryos. (**c**) EGFP assay of embryos injected with morpholino and GFP-pri-miR-205 construct. The miRNA activity is decreased upon morpholino co-injection, resulting in an increased EGFP signal. Image adapted with permission from Kloosterman WP et al. (2007) PLoS Biol

the morpholinos directly targeting the mature miR-375 only slightly decreased the expression of miR-375; however the morpholino was able to inhibit the function of miR-375 in a GFP reporter assay.

The knockdown of various miRNAs in zebra fish embryos by morpholinos did not result in gross morphological deformities, showing no significant effect on embryonic development after 4 days. While there were initially no developmental defects 5 days after morpholino-mediated knockdown of miR-375, 7 days after injection a majority of the embryos died. Further investigation revealed that the morpholino-injected embryos displayed scattered islet cells resulting in drastic changes in insulin expression. Close monitoring of morpholino-injected embryos showed that loss of miR-375 leads to malformation of the endocrine pancreas in zebra fish. Thus, morpholinos can selectively target mature miRNA sequences or be used to inhibit Drosha or Dicer cleavage through targeting of precursor molecules. This efficient technique allows for reliable knockdown of various miRNAs during the early stages of zebra fish development. This study demonstrates the importance of miRNA maturation in vivo and that antisense agents are capable of targeting not only the function of mature miRNA but also the processing by both Dicer and Drosha enzymes.

3.3 Inhibition of Pre-miRNA In Vitro Without Assessing Specificity

Based on the convergence of the miRNA biogenesis pathway on the same set of processing enzymes irrespective of the RNA, nonspecific targeting of miRNA maturation and function is relatively easier to achieve than the previously discussed specific targeting. Consequently, there are many more examples of both small-molecule and biomacromolecular modulators of the processing of pre-miRNA. This is especially true in vitro where direct assays have been developed with the isolated components of the maturation pathway to ensure modulation of their function by the effector molecule.

3.3.1 Inhibition of Pre-miRNA Maturation by Peptides

One such in vitro assay has been developed in the Arenz laboratory to monitor the ability of Dicer to process pre-miRNA into its active mature miRNA analog [53, 54]. The fluorescence-based assay was developed using a pre-let-7 probe due to its prevalence in miRNA studies and its well-defined Dicer-mediated cleavage. This probe was modified to contain a 5′ fluorescein label and a 3′ dabcyl quencher (Fig. 8), affording fluorescence quenching when the properly folded hairpin pre-miRNA is formed; however, upon Dicer maturation of the pre-miRNA the proximity of the fluorophore and the quencher is lost, resulting in a fluorescence signal. Incubation of the probe with either recombinant Dicer or cellular lysate containing Dicer in vitro resulted in a three- to tenfold increase in fluorescence. With the assay firmly established, inhibitors could be screened for a decrease in fluorescence signal corresponding to the prevention of pre-miRNA Dicer-mediated processing. In order to ensure inhibition, three dodecapeptides derived from the amino acid sequence of Dicer and kanamycin (a known binder of RNA) were tested for modulation of pre-miRNA maturation. At high concentrations of 100 μM, one peptide was able to inhibit fluorescence approximately 85 %, while kanamycin exhibited a 40 % reduction. This effect was reduced at lower concentrations and when using cell lysate Dicer (due to nonspecific binding to cellular nucleic acids) but was still observable at 10–30 %. Based on the initial peptides used in the assay, the

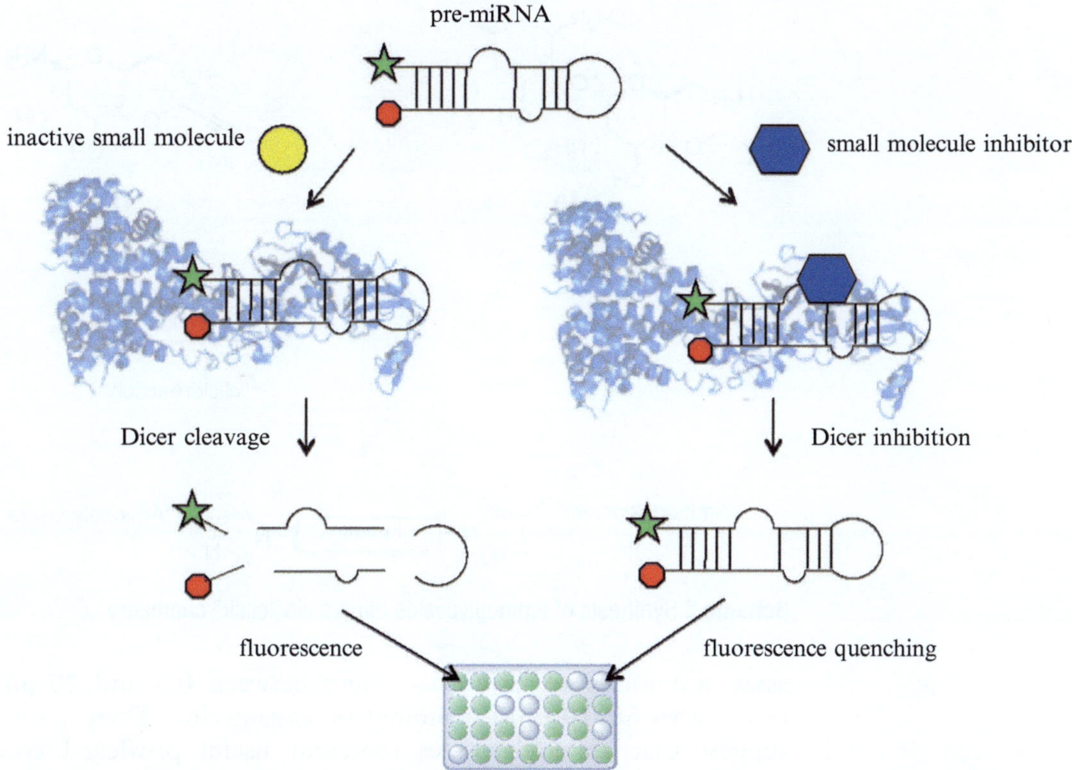

Fig. 8 Pre-miRNA maturation assay for Dicer processing. A dual-labeled pre-miRNA with a fluorescent molecule and a quencher in close proximity is incubated with Dicer in vitro. Small-molecule inhibition of processing prevents RNA cleavage and maintains fluorescence quenching, while inactive molecules result in Dicer processing and a fluorescent signal

disruption of pre-miRNA processing is most likely nonspecific; however, by screening against large-peptide or small-molecule libraries with different pre-miRNA reporters it may be feasible to identify specific inhibitors of the processing event. Either way, the targeting of Dicer-mediated maturation represents an excellent target in decreasing the intracellular concentrations of mature miRNAs.

3.3.2 Inhibition of Pre-miRNA Maturation by Aminoglycosides

Based on the well-documented ability of aminoglycosides to target and bind RNA, they become a logical source for pre-miRNA binders to modulate maturation. As noted in the previous assay kanamycin was able to elicit some effect on Dicer processing of pre-let-7 miRNA, suggesting the viability of this class of compounds as inhibitors. Towards this end a series of aminoglycoside derivatives based on 2-deoxystreptamine (**11**) and neamine (**10**) derivatives were prepared and dimerized via 1,3-dipolar cycloaddition reactions to increase the valency and thus potency of their RNA binding activity (Scheme 2) [55]. Six conjugates of **10** and **11** were assayed in the previously described fluorescence quenching Dicer

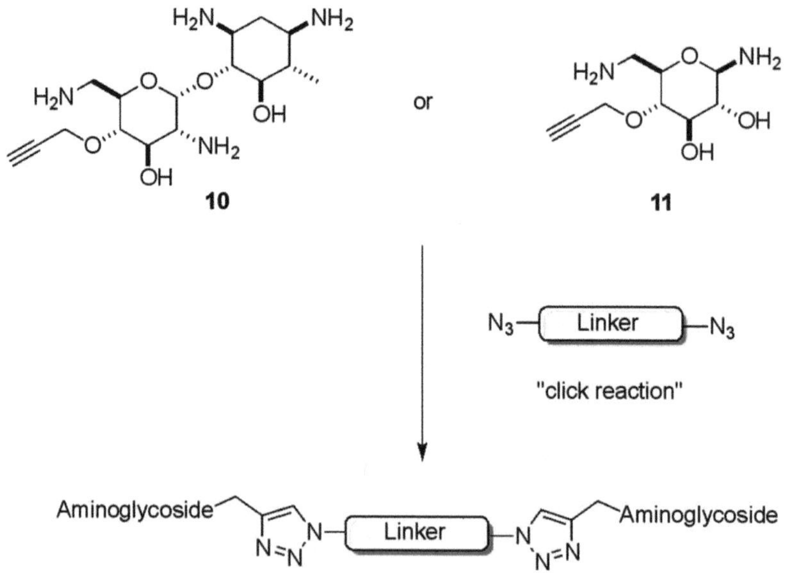

Scheme 2 Synthesis of aminoglycoside dimers via "click" chemistry

assay and found to have IC_{50} values between 0.5 and 20 μM (relative to 50–100 μM exhibited by kanamycin). These results suggest that aminoglycosides represent useful privileged core structures for the inhibition of miRNA maturation. However, the specificity of these molecules has yet to be investigated, and increasing their affinity for general RNA binding may lead to issues within a cellular environment.

These aminoglycoside inhibitors were then employed in the development of a more robust assay for the measurement of pre-miRNA maturation based on branched rolling-circle-amplification (BRCA) [56]. Though a variety of assays have been developed to monitor the maturation of miRNA, especially in the presence of inhibitors, current methods contain many functional limitations. Previously described [32] P-labeled pre-miRNA PAGE assays to evaluate miRNA maturation inhibition are tedious and unsuitable for high-throughput screenings. Comparatively, BRCA allows for sensitive quantifications of miRNA levels. This particular method utilizes a cyclic DNA template with a complementary primer that through polymerase binding triggers the formation of single-stranded DNA product. The single-stranded DNA then serves as a template for further amplification with a secondary primer. Using this technique, Arenz et al. designed a label-free BRCA-based assay that could selectively detect the formation of mature miRNAs.

In order to engineer discrimination between pre-miRNA and mature miRNA, the circular DNA template was constructed so that only mature miRNA could initiate BRCA. The 3′ end of mature miRNA, which is inaccessible in the pre-miRNA, was used for BRCA elongation (Fig. 9). After designing a 65-mer circular

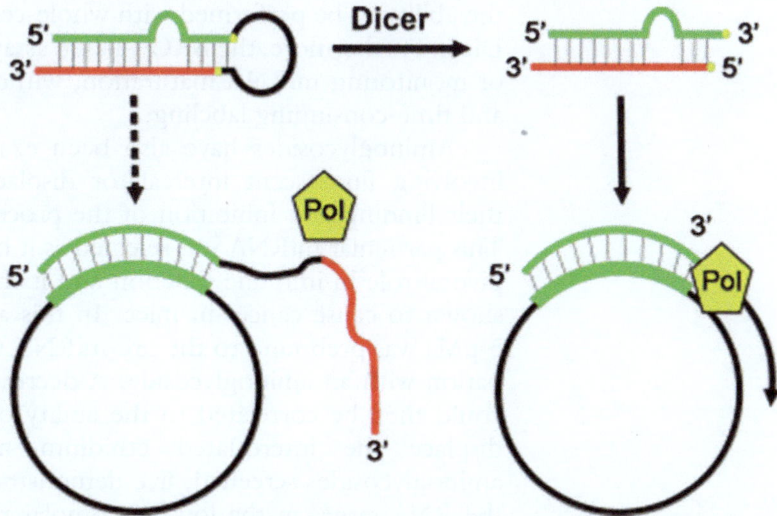

Fig. 9 Branched rolling-circle-amplification (BRCA) assay for detection of pre-miRNA maturation inhibitors. If the pre-miRNA is not fully processed, the residual hairpin blocks polymerase amplification of the circular DNA template, whereas if pre-miRNA is fully matured exponential amplification of the template can occur. Image adapted with permission from Neubacher S et al. (2011) Chem Bio Chem

DNA template with a sequence complementary to mature let-7 miRNA, real-time PCR was used to monitor the BRCA reactions. When two polymerases, Kle^{-exo} and Bst, were used in the presence of the control, untreated pre-let-7, BRCA reactions were not observed. Conversely, when the pre-let-7 template was preincubated with recombinant Dicer, there was significant DNA polymerization. The Dicer cleavage and BRCA reaction could be conducted in an efficient one-pot reaction due to the reaction optimal temperatures for Kle^{-exo} and Dicer. However, due to the 40-fold increased signal amplification Bst was selected to optimize the reaction despite requiring a higher temperature. To measure the selectivity for precursor or mature miRNA, pre-let-7 was incubated with Dicer for increasing periods of time. The results indicated a strong dependence on Dicer incubation time, making this assay particularly useful for measuring miRNA maturation. Moreover, the assay can also selectively monitor miRNA maturation in the presence of multiple miRNAs.

The BRCA-based assay was further tested to determine the effects of the known aminoglycoside inhibitors of Dicer cleavage. A control BRCA reaction was designed in order to exclude potential false hits that could occur if the inhibitor altered polymerization patterns. The results of these assays yielded similar inhibition values as determined by the previously described fluorescence-based assays, confirming its validity as a screen. However, this method provides several benefits over other techniques, including the ability to be performed with whole-cell lysates or recombinant Dicer. Furthermore, the BRCA-based assay provides a useful means of monitoring miRNA maturation, without the use of expensive and time-consuming labeling.

Aminoglycosides have also been examined in another assay involving fluorescent intercalator displacement (FID) to assess their binding and inhibition of the processing of miR-151 [57]. This particular miRNA was selected as it has been shown to have a pivotal role in immune function and its high expression has been shown to cause cancer in mice. In this assay ethidium (RNA K_d 5 μM) was prebound to the pre-miRNA target, followed by incubation with an aminoglycoside. A decrease in fluorescence signal could then be correlated to the ability of the aminoglycoside to displace the intercalated ethidium molecule. Of the ten aminoglycosides screened, five demonstrated moderate affinity to the RNA target in the low micromolar range. To confirm actual binding of the molecules, rather than a simple structural alternation of the RNA target, direct binding studies were conducted via probing the thermal melting temperature of the pre-miRNA target. An increase in melting temperature corresponds to a binding interaction between the RNA and the aminoglycoside, leading to increased stabilization of the complex. The data suggests that aminoglycosides in particular are effective in binding to the pre-miRNA target and do so in a dose–response relationship with ligand concentration.

When examined in a Dicer-catalyzed pre-miRNA processing SDS-PAGE assay, an interesting effect was observed, as only three of the aminoglycosides exhibited any inhibitory effect at all (~10 %) despite binding the pre-miRNA. Additionally, the assay demonstrated that degradation products corresponding to nonspecific Dicer processing were observed and were also inhibited by the aminoglycosides. This indicates that the aminoglycosides are not only nonspecific inhibitors but also nonspecific binders. This provides a cautionary point that demonstrates that RNA binding is not necessarily correlated to the inhibition of pre-miRNA maturation. Consequently, a higher degree of specificity and inhibitory activity is a requisite before any of these compounds can have therapeutic relevance.

3.3.3 Inhibition of Pre-miRNA Maturation by Other Small Molecules

In the previously described FID assay by Herdewijn et al., other small-molecule RNA binders were examined, most notably various oligonucleotide intercalators [57]. Many of these molecules, including DAPI, proflavine, doxorubicin, and Hoechst 33258, exhibited high levels of both binding and inhibition by SDS-PAGE analysis of Dicer-mediated pre-miRNA processing. However, these compounds were found to be neither specific binders nor specific inhibitors, as nonspecific Dicer cuts were also inhibited. It is speculated that potential hybrids of these intercalating compounds with

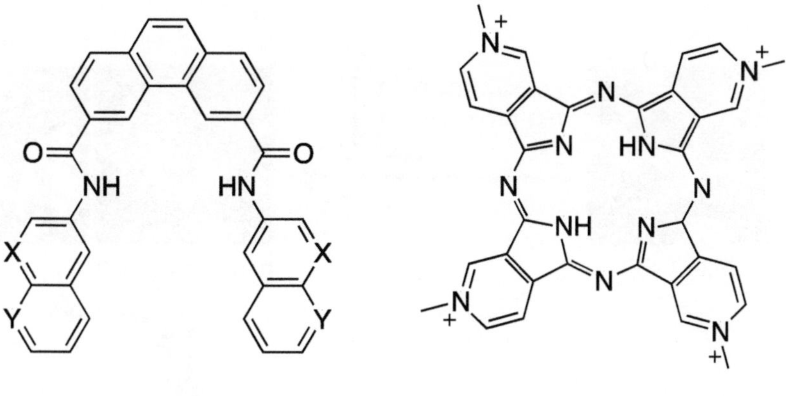

Fig. 10 Quadruplex binding compounds found to inhibit maturation of shRNA

aminoglycosides may afford a heightened degree of specificity and inhibition of certain pre-miRNAs.

As a means of achieving some form of specificity of pre-miRNA processing, work in the Hartig laboratory utilized known G-quadruplex small-molecule binders [58]. While this study was conducted using short hairpin RNAs (shRNAs) (an analog of pre-miRNAs in the siRNA pathway), specific G-rich RNA sequences were able to bind bisquinolinium (**12**) and porphyrazine (**13**) compounds with 20–400 nM dissociation constants and inhibit Dicer processing as demonstrated via SDS-PAGE analysis (Fig. 10). While this method is not amenable to high-throughput screening and some knowledge of the pre-miRNA secondary is required, it does represent a means of obtaining some specificity for G-rich sequences in pre-miRNAs. Unfortunately, when translated to a mammalian cell system the presence of the small molecules was not able to reverse the silencing effects of the shRNAs on a reporter gene. This is potentially the result of poor cellular uptake of the compounds or inability to discriminate between the desired shRNAs and various other G-quadruplexes present within the cell. Thus, in order to develop practical compounds, a significantly higher degree of specificity must be achieved for the targeted pre-miRNA.

3.4 Modulation of Pre-miRNA In Vivo Non-specifically

Recently, some classes of small molecules have also been implicated in the upregulation of RNA interference pathways by increasing Dicer-mediated processing. By developing a cell-based screening assay, Shan et al. identified the small molecule enoxacin (**14**) as a promoter of miRNA biogenesis [59]. Researchers developed an assay in HEK293T cells that stably expressed both a GFP reporter and an shRNA. The shRNA required maturation by Dicer to yield a siRNA that specifically targets the GFP mRNA, resulting in a reduction of fluorescence. Using this reporter system to screen

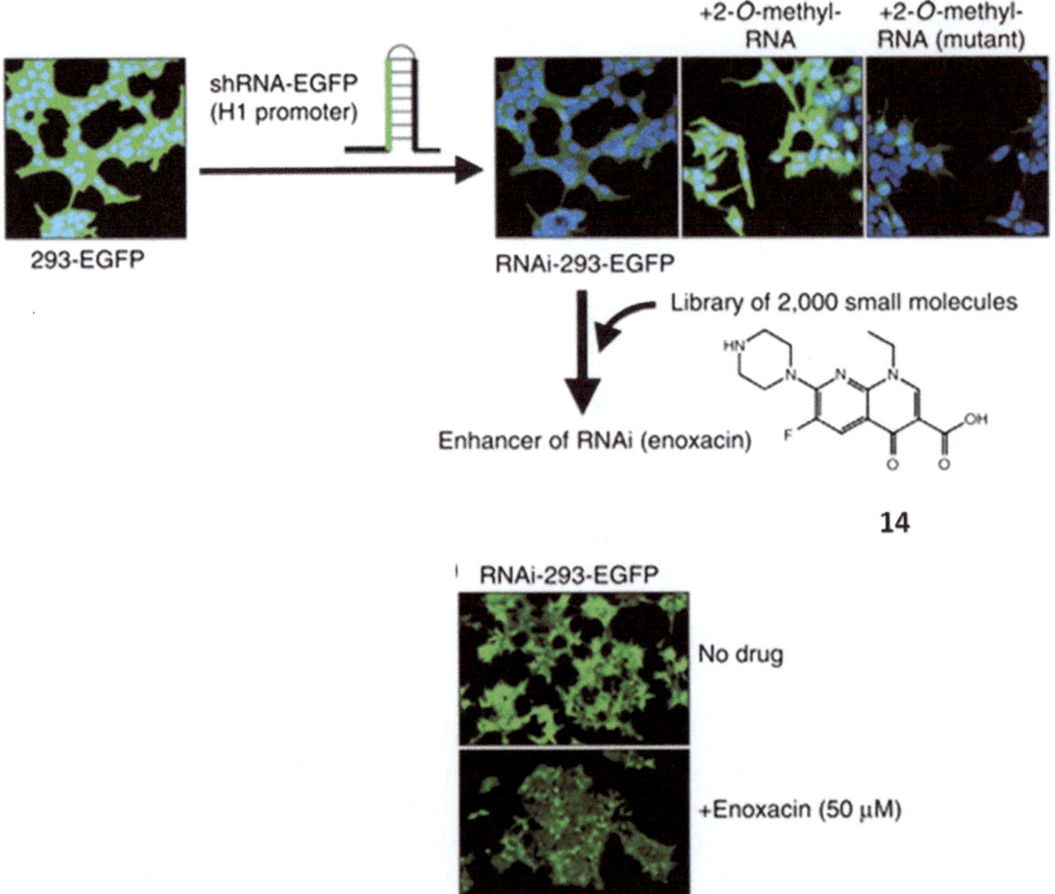

Fig. 11 EGFP-based in vivo assay for the elucidation of miRNA/siRNA pathway activators. Cells expressing EGFP were treated with shRNA targeting the gene and an array of small molecules. Enoxacin was found to increase the degree of EGFP silencing relative to a non-treated control, signifying an activator of the RNAi pathway. Image adapted with permission from Shan G et al. (2008) Nat Biotechnol

2,000 FDA-approved compounds, **14** was found to enhance siRNA-mediated degradation of mRNA as GFP silencing was higher than the cellular controls (Fig. 11). Interestingly, the Xi laboratory also developed a similar assay and independently discovered the same compound as a positive effector [60].

Though other structurally similar molecules were also screened, enoxacin was the most effective at enhancing RNAi activity. Microarray analyses demonstrated that enoxacin-treated cells did not display significant changes in gene expression, indicating that enoxacin likely affects shRNA/pre-miRNA maturation. Northern blot analysis showed that the presence of enoxacin increased the levels of both pre-let-7 and pre-miR-30 only in the presence of both Dicer and TRBP (but not Dicer alone). This trend suggests that enoxacin likely aids in the loading of RNA duplexes onto RISCs, resulting in enhanced RNAi activity. Moreover, TRBP knockdown suggested that enoxacin activity is TRBP dependent in vivo.

To better determine the action of enoxacin in vivo, researchers utilized a similar shGFP/GFP reporter assay in transgenic mice. Enoxacin was injected, and after incubation the mRNA levels of GFP were determined. Similar to the in vitro studies, the addition of enoxacin increased the knockdown of GFP by up to 60 %. The use of small molecules such as enoxacin to regulate the RNAi pathway may provide important information about the cellular effects of miRNA biogenesis. This exciting result indicates that it is possible to also upregulate the effect of miRNAs in cells and that it is possible to recapitulate cellular studies and the small molecule effect in model organisms.

The Jeang laboratory reported another example of a small-molecule modulator of miRNA maturation [61]. Using cells transfected with a dual-luciferase reporter assay and an shRNA targeting firefly luciferase, 530 small molecules were screened for the ability to inhibit the knockdown of firefly luciferase relative to the *Renilla* control. Two compounds, polylysine and trypaflavine, were found to reproducibly affect the luciferase ratios relative to a non-treated control. After several elegant experiments to elucidate the mechanism of action, polylysine was found to inhibit the formation of Dicer–RNA complexes, effectively inhibiting the last maturation step in the biogenesis pathway. Alternatively, trypaflavine was found to block miRNA loading into the RISC complex by inhibiting either TRBP/AGO2 or RHA/AGO2 associations. Even more excitingly, the efficacy of these compounds was examined in a mouse model where cancerous cells over-expressing miR-93 and miR-130 were implanted in a mouse leading to tumor formation. Treatment with either compound resulted in a significantly reduced ability of the cells to promote tumor formation (Fig. 12). This result demonstrates that it is possible to have therapeutic compounds that alter miRNA maturation in a nonspecific fashion. While various other small molecule approaches to miRNA/siRNA regulation have been reported, most focus on the modulation of miRNA function and not necessarily maturation and thus will not be discussed. Nonetheless, this represents an exciting and rapidly developing field for the development of therapeutic small-molecule modulators of miRNAs.

4 Manipulation of miRNA Function in Living Cells with Oligonucleotides

Due to the importance of miRNAs in multiple disease pathologies, various studies focus on developing loss-of-function models in vivo. The most direct route to achieve these knockdowns is via the use of complementary oligonucleotides which directly base pair to

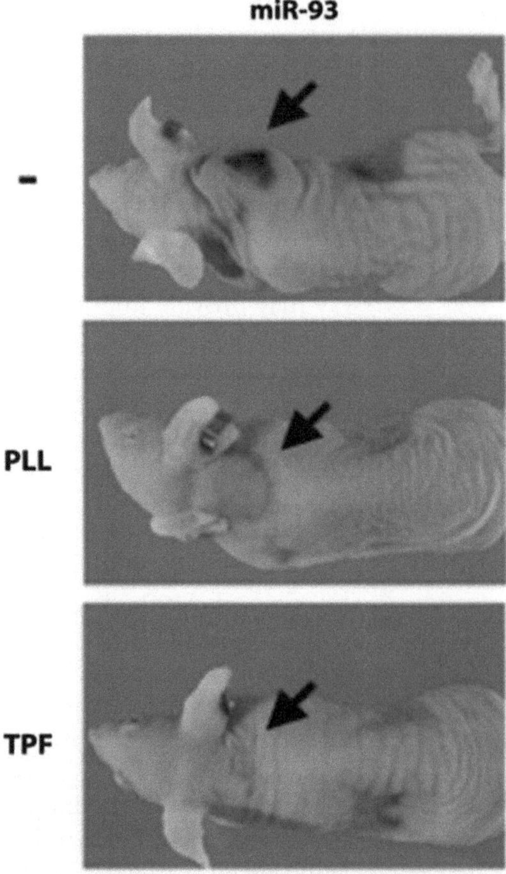

Fig. 12 Therapeutic relevance of miRNA biogenesis inhibitors. A tumor mouse model was treated with either polylysine (PLL) or trypaflavine (TPF), which were previously found to inhibit miRNA function/maturation. Only drug-treated mice exhibited a reduction in tumor growth demonstrating the in vivo efficacy of the small-molecule inhibitors. Image adapted with permission from Watashi K et al. (2010) J Biol Chem

the mature miRNAs, resulting in a loss-of-function phenotype. While this does not directly affect miRNA maturation, its prevalence within the field is worth noting.

Early studies employed 2′-O-methyl oligonucleotides directed against let-7 miRNA and demonstrated efficient and irreversible base pairing to the miRNA guide strand within RISC [23]. Additionally, they helped elucidate key protein components of RISC within *C. elegans* and suggested the ability of these miRNA-targeting oligonucleotides to achieve a let-7 loss-of-function phenotype within a model organism by simple injection into the larvae. Finally, the antisense oligonucleotides were able to show that RISC function was significantly more impacted by antisense binding to RISC rather than binding to the complementary mRNA sequence, suggesting an active mechanism of gene silencing by RISC.

Another approach by the Stoffel laboratory utilized chemically engineered single-stranded RNAs, known as antagomirs, that allow for efficient and specific silencing of various miRNAs within a cell or a model organism [24]. These synthetic analogues consisted of a cholesterol support fused to single-stranded RNA that was perfectly complementary to a specific miRNA. Gratifyingly, intravenous administration of an antagomir significantly decreased the level of its corresponding miRNA in various mouse tissues. Since many miRNAs are confined to certain types of tissues, such as the liver-specific miRNA-122, antagomirs may be therapeutically useful in selectively silencing endogenous miRNAs.

To this end, various antagomirs were synthesized, including the miR-122-specific antagomir-122. Though partially and fully modified phosphorothioate RNA strands were also studied, the cholesterol-conjugated antagomirs were found to be the most effective and specific in silencing miRNA. Administration of antagomir-122 in mice resulted in miR-122 degradation products as well as significantly decreased levels of miR-122, indicating that antagomir-122 silences miR-122 through degradation.

This pharmacological silencing further elucidated the role of miR-122 in the regulation of many mRNAs that contributed to the liver phenotype. Antagomir-122-treated mice displayed specific phenotypic changes, such as decreased plasma cholesterol levels, linking miR-122 to the regulation of cholesterol biosynthesis. Gene analysis from antagomir-treated mice identified recognition motifs within genes that interacted with miR-122, leading to upregulation or downregulation of multiple genes. The recognition motif, or miR-122 nucleus sequence, was the most over-represented hexamer found in a bioinformatics analysis. When the miR-122 nucleus was incorporated into a luciferase reporter system, co-transfection with miR-122 led to a significantly repressed reporter when compared to control miRNA transfections. Thus binding at the miR-122 nucleus directly represses mRNA expression.

Antagomirs have been proven to provide a powerful means to assess gene regulation in vivo. Silencing of select endogenous miRNAs or a combination of miRNAs within mice leads to a better understanding of the function and roles of various miRNAs. Indeed, the long-lasting activity of antagomirs coupled with broad biodistribution and low toxicity makes them ideal candidates for therapies aimed at diseases linked to miRNA misregulation.

Though antagomirs provide an efficient method to target miRNAs, they require extensive sequence complementarity, and the requisite pairing of one oligonucleotide per mature miRNA often obviates high doses in order to obtain an in vivo effect. Comparatively, an over-expressed miRNA decoy target can similarly silence its cognate miRNA by acting as a competitive inhibitor [62]. This decoy may also be able to interact with specific sequences that are shared by members of the same miRNA family. Ebert et al.

investigated this hypothesis by constructing various decoys, termed microRNA sponges, that contain tandem miRNA-binding sites. These transcripts were inserted into the 3′ UTR of a destabilized GFP gene (used to measure transformation efficiency) driven by either a CMV promoter or a U6 small nuclear RNA promoter–terminator vector. The RNA sequence of the sponges was then modified to be either perfectly or imperfectly complementary to the miRNA of interest.

Expression assays with *Renilla* luciferase genes regulated by miR-20 revealed an increase in relative target expression after co-transfection with either CMV or U6 sponges. Both types of sponges inhibited miR-20, rescuing the expression of the luciferase target, with the imperfectly matched sponges outperforming the perfectly complementary sponges. The ability of miRNA sponges to inhibit endogenous miRNA was compared to previously developed antisense therapies by transfecting with target reporters. While most of the oligonucleotides depressed the target reporter expression level, the most significant change was observed with the miRNA sponges (Fig. 13). The combined use of other oligonucleotides and miRNA sponges resulted in the largest recovery of target expression levels.

Another benefit of miRNA sponges is the ability of a single sponge to inhibit miRNA families that share a common heptameric seed. A target reporter with binding sites for miR-30c was used to verify the seed specificity of a miRNA sponge against miR-30e compared to a 2′-O-methyl antisense oligonucleotide against miR-30e. The sequences for miR-30e and miR-30c have the same heptameric seed; however the antisense oligonucleotide only slightly depressed the target miR-30c. The miR-30e sponge restored target expression over fourfold, indicating cross-reactivity with multiple miR-30 family members.

While antisense oligonucleotides provide an efficient means to inhibit endogenous miRNAs, miRNA sponges act as efficient decoys that can target multiple miRNA family members. Future work with miRNA sponges aims at increasing their potency, perhaps through additional seed-binding sites and modulation of their complementarity.

Due to the gene regulatory role of miRNAs, especially in developmental processes, the immediate knockdown of miRNA function upon introduction of an antisense agent may not be optimal and result in undesired side effects. Towards this end the Li laboratory developed photoactivatable antisense oligonucleotides that remain inactive in their miRNA silencing until a brief irradiation with 365 nm light [63]. This affords a degree of spatial and temporal control over the antisense reagent, allowing for complex investigations into the role of miRNAs in developmental processes. Specifically, a blocking sequence complementary to the antisense reagent was intramolecularly attached to the 2′-O-methyl antisense

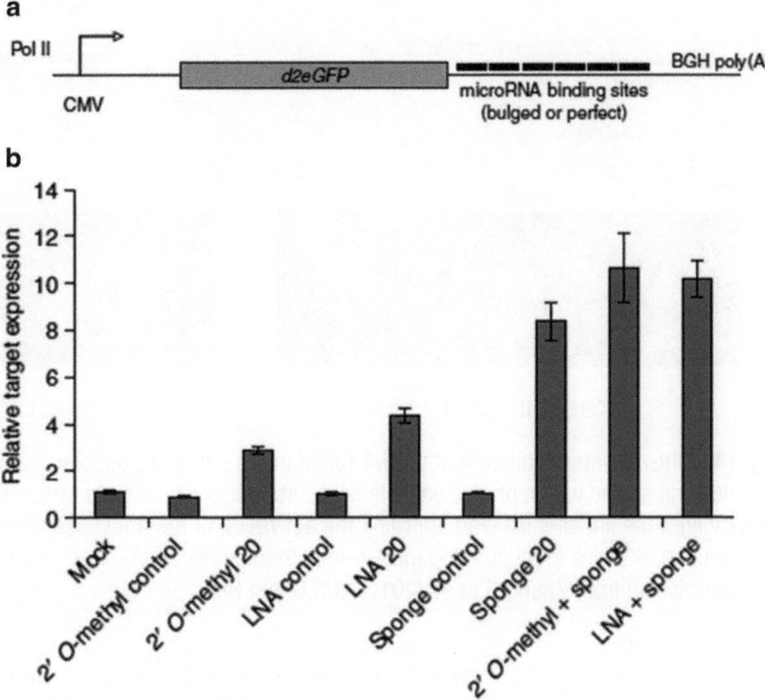

Fig. 13 Increased efficiency of miRNA sponges relative to traditional targeting of miRNAs. (**a**) miRNA sponge construct containing several miRNA-binding sites specific for miRNA seed family that were either exactly complementary or bulged. These constructs were co-transfected with a luciferase reporter plasmid with the complementary to the sponge sites. (**b**) Luciferase target suppression using a variety of miRNA inhibitory oligonucleotide strategies. Co-transfection with sponge constructs resulted in a significantly higher degree of miRNA inhibition than obtained with traditional oligonucleotide inhibitors. Image adapted with permission from Ebert MS et al. (2007) Nat Met

oligonucleotide with a fluorescent and photoreactive linker abrogating its silencing ability. The lys-6 miRNA in *C. elegans* was selected as a target due to its role in right/left asymmetric differentiation in the nervous system (Fig. 14). After optimization of the length of the blocking oligonucleotide, the antisense agents were injected into the gonad of adult worms, and miRNA activity was knocked down at various time points during larval development to assess when lys-6 miRNA was active in nervous system development. This approach was successful, as inactivation before comma phase led to a complete knockout phenotype, while activation at any timepoint after the comma phase had no phenotypic effect, demonstrating that lys-6 is only truly essential in early development. This extension of antisense technology is useful in illustrating the importance of miRNAs not only in diseases but also within a developmental context.

A recent advancement in the silencing of miRNA in vitro and in vivo includes the integration of synthetic DNA strands with

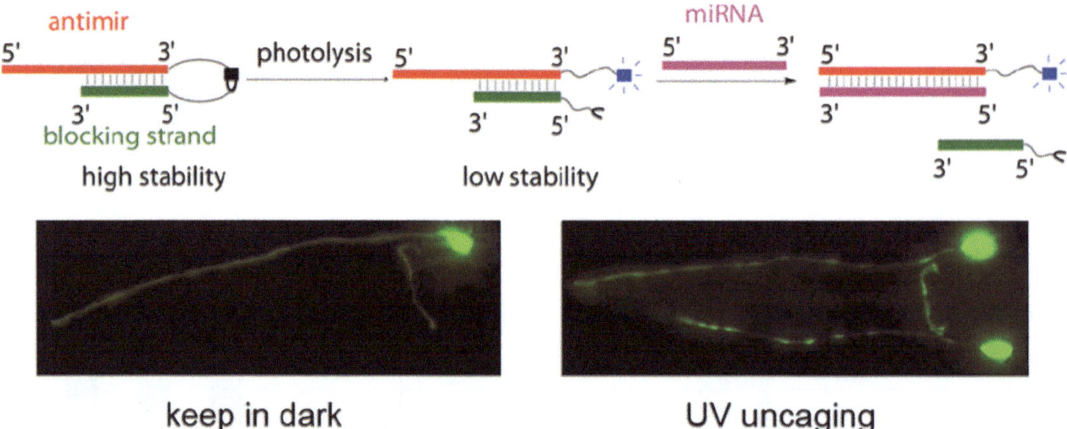

Fig. 14 Light-activated inhibition of miRNA function. An inhibitory oligonucleotide was intramolecularly linked to a blocking strand with a photocleavable linker, inhibiting its ability to target miRNAs. Upon a brief irradiation with UV light the linker is cleaved affording the activation of the antisense oligonucleotide. This construct was proven to be effective in modulating miRNA-mediated developmental processes in *C. elegans*. Image adapted with permission from Zheng G et al. (2012) ACS Chem Biol

catalytic ribozyme motifs. These oligonucleotide enzymes are capable of cleaving RNA substrates in a sequence-specific fashion leading to gene inactivation. DNAzymes are composed of a conserved catalytic core with two flanking regions that are capable of binding an RNA-substrate sequence [64]. The binding and cleaving actions of DNAzymes prove to be efficient and specific inhibitors of miRNA, with locked nucleic acid (LNA) modifications enhancing the catalytic activity. This new class of antisense inhibitors has been coined "antagomirzymes" [65].

Through in vitro assays antagomirzymes were found to cleave specific miRNA targets with approximately 70 % of the miRNA targets cleaved by the LNA-based analogs. The catalytic activity was confirmed via disruption of the catalytic domain, which eradicated the ability of the antagomirzyme to inhibit miRNAs. Moreover, an in vivo luciferase assay examined hsa-miR-372 and hsa-miR-373 targets, with LNA antagomirzymes restoring the reporter level by up to 80 %. In comparison, oligonucleotides with a shuffled catalytic domain led to suppressed expression, again highlighting the catalytic action of the antigomirzyme rather than simply its binding ability. While both forms of antigomirzymes effectively cleave miRNAs in vitro and in vivo, very high concentrations are required to be therapeutically useful. Future work in this area requires the identification of better cleavage sites within miRNA targets and optimization of cellular delivery.

A final mechanism to regulate RNAi within mammalian cells involves a hybrid approach that engineers an RNA aptamer onto an shRNA in order to modulate miRNA levels using a small-molecule effector [66]. Researchers constructed an shRNA (targeting

EGFP) fused to an RNA aptamer evolved for the small molecule theophylline. A control construct was also prepared which lacked the aptamer motif on the shRNA. Monitoring of the GFP fluorescence was correlated to the ability of the shRNA to silence gene expression. The addition of theophylline led to the dose-dependent increase in fluorescence for the shRNA fused to the RNA aptamer; however, the theophylline had no effect on the control shRNA. RNAi inhibition by theophylline recovered the EGFP signal, and a similar result was also obtained when targeting the DsRed gene.

A variety of constructs were developed in order to elucidate theophylline's mechanism of action. The results indicated that theophylline's interaction with the RNA aptamer inhibits Dicer cleavage likely due to the close proximity of the binding pocket to the Dicer cleavage site. To observe the inhibition of siRNA in vivo, researchers co-transfected an EGFP and an shRNA expression vector and then incubated cells in the presence or the absence of 10 mM theophylline. Northern blot analysis revealed that siRNA production was almost completely inhibited when the RNA aptamer and theophylline were both present. Conversely, there was no significant change in the siRNA levels for the vector lacking the theophylline binding aptamer. Though multiple aptamer constructs were prepared, the highest degree of inhibition was observed when the putative Dicer cleavage site overlapped with the theophylline-binding site.

Though multiple methods have aimed to use small molecules as a means to regulate RNAi pathways, they involve careful design and engineering. The aptamer-fused shRNA system may be used in combination with other regulatory systems that also target the RNAi pathway. The benefit of this particular aptamer-fused approach is a faster, more direct response than with transcriptional control achieved by targeting the Dicer-mediated cleavage. However, to be more applicable in the targeting of miRNAs, this methodology must be adapted to produce shRNAs targeting miRNAs directly, and other approaches previously described appear to be somewhat more straightforward. Yet, one significant advantage is the ability to modulate miRNA function at a given time via introduction of the small-molecule effector.

5 Conclusions

The ever-increasing relevance of miRNAs within biological systems requires the development of new tools to better understand and modulate its function. Specifically, the ability to modulate its maturation using either small molecule or biomacromolecule approaches has a plethora of applications including the development of novel therapeutics for multiple diseases and disorders. While RNA has traditionally been considered a "non-druggable" target many innovations have recently been made within the field to make targeting of RNA sequences a reality. This can be achieved in either a specific or a nonspecific fashion, albeit distinct advantages and disadvantages are associated with either approach. Moreover, the selection of targeting agent (e.g., small molecule, peptide, oligonucleotide) is also very diverse and should be fine-tuned for the desired application. Overall, the modulation of miRNA maturation is rapidly advancing, providing new insights and therapeutic agents every year. This progress is poised only to accelerate as new tools and understandings further demonstrate the importance of miRNAs.

References

1. Hannon GJ (2002) RNA interference. Nature 418(6894):244–251
2. Grishok A, Pasquinelli AE, Conte D, Li N, Parrish S, Ha I, Baillie DL, Fire A, Ruvkun G, Mello CC (2001) Genes and mechanisms related to RNA interference regulate expression of the small temporal RNAs that control C. elegans developmental timing. Cell 106(1): 23–34
3. Mattick JS (2001) Non-coding RNAs: the architects of eukaryotic complexity. EMBO Rep 2(11):986–991
4. Deiters A (2010) Small molecule modifiers of the microRNA and RNA interference pathway. AAPS J 12(1):51–60
5. Georgianna WE, Young DD (2011) Development and utilization of non-coding RNA-small molecule interactions. Org Biomol Chem 9(23):7969–7978
6. Mattick JS (2009) Deconstructing the dogma: a new view of the evolution and genetic programming of complex organisms. Ann N Y Acad Sci 1178:29–46
7. Mattick JS (2010) The central role of RNA in the genetic programming of complex organisms. An Acad Bras Cienc 82(4):933–939
8. Sun W, Julie Li YS, Huang HD, Shyy JY, Chien S (2010) microRNA: a master regulator of cellular processes for bioengineering systems. Annu Rev Biomed Eng 12:1–27
9. Bartel DP (2009) MicroRNAs: target recognition and regulatory functions. Cell 136(2): 215–233
10. He L, Hannon GJ (2004) MicroRNAs: small RNAs with a big role in gene regulation. Nat Rev Genet 5(7):522–531
11. Kwak PB, Iwasaki S, Tomari Y (2010) The microRNA pathway and cancer. Cancer Sci 101(11):2309–2315
12. Ambros V, Chen X (2007) The regulation of genes and genomes by small RNAs. Development 134(9):1635–1641
13. Garzon R, Marcucci G, Croce CM (2010) Targeting microRNAs in cancer: rationale, strategies and challenges. Nat Rev Drug Discov 9(10):775–789
14. Thomson JM, Newman M, Parker JS, Morin-Kensicki EM, Wright T, Hammond SM (2006) Extensive post-transcriptional regulation of microRNAs and its implications for cancer. Genes Dev 20(16):2202–2207
15. Czech B, Malone CD, Zhou R, Stark A, Schlingeheyde C, Dus M, Perrimon N, Kellis M, Wohlschlegel JA, Sachidanandam R, Hannon GJ, Brennecke J (2008) An endogenous small interfering RNA pathway in Drosophila. Nature 453(7196):798–802
16. Ghildiyal M, Zamore PD (2009) Small silencing RNAs: an expanding universe. Nat Rev Genet 10(2):94–108
17. Okamura K, Hagen JW, Duan H, Tyler DM, Lai EC (2007) The mirtron pathway generates microRNA-class regulatory RNAs in Drosophila. Cell 130(1):89–100
18. Inose H, Ochi H, Kimura A, Fujita K, Xu R, Sato S, Iwasaki M, Sunamura S, Takeuchi Y, Fukumoto S, Saito K, Nakamura T, Siomi H, Ito H, Arai Y, Shinomiya K, Takeda S (2009) A microRNA regulatory mechanism of osteoblast differentiation. Proc Natl Acad Sci U S A 106(49):20794–20799
19. Tsai MC, Spitale RC, Chang HY (2011) Long intergenic noncoding RNAs: new links in cancer progression. Cancer Res 71(1):3–7
20. Weinberg MS, Wood MJ (2009) Short non-coding RNA biology and neurodegenerative disorders: novel disease targets and therapeutics. Hum Mol Genet 18(R1):R27–R39
21. Kloosterman WP, Plasterk RH (2006) The diverse functions of microRNAs in animal development and disease. Dev Cell 11(4): 441–450
22. Lu J, Getz G, Miska EA, Alvarez-Saavedra E, Lamb J, Peck D, Sweet-Cordero A, Ebert BL,

Mak RH, Ferrando AA, Downing JR, Jacks T, Horvitz HR, Golub TR (2005) MicroRNA expression profiles classify human cancers. Nature 435(7043):834–838

23. Hutvágner G, Simard MJ, Mello CC, Zamore PD (2004) Sequence-specific inhibition of small RNA function. PLoS Biol 2(4):E98
24. Krützfeldt J, Rajewsky N, Braich R, Rajeev KG, Tuschl T, Manoharan M, Stoffel M (2005) Silencing of microRNAs in vivo with 'antagomirs'. Nature 438(7068):685–689
25. Calin GA, Cimmino A, Fabbri M, Ferracin M, Wojcik SE, Shimizu M, Taccioli C, Zanesi N, Garzon R, Aqeilan RI, Alder H, Volinia S, Rassenti L, Liu X, Liu CG, Kipps TJ, Negrini M, Croce CM (2008) MiR-15a and miR-16-1 cluster functions in human leukemia. Proc Natl Acad Sci U S A 105(13):5166–5171
26. Costa FF (2009) Non-coding RNAs and new opportunities for the private sector. Drug Discov Today 14(9–10):446–452
27. Li Y, He C, Jin P (2010) Emergence of chemical biology approaches to the RNAi/miRNA pathway. Chem Biol 17(6):584–589
28. Thomas JR, Hergenrother PJ (2008) Targeting RNA with small molecules. Chem Rev 108(4): 1171–1224
29. Zhang S, Chen L, Jung EJ, Calin GA (2010) Targeting microRNAs with small molecules: from dream to reality. Clin Pharmacol Ther 87(6):754–758
30. Du T, Zamore PD (2005) microPrimer: the biogenesis and function of microRNA. Development 132(21):4645–4652
31. Lee Y, Ahn C, Han J, Choi H, Kim J, Yim J, Lee J, Provost P, Rådmark O, Kim S, Kim VN (2003) The nuclear RNase III Drosha initiates microRNA processing. Nature 425(6956): 415–419
32. Denli AM, Tops BB, Plasterk RH, Ketting RF, Hannon GJ (2004) Processing of primary microRNAs by the Microprocessor complex. Nature 432(7014):231–235
33. Hutvágner G, McLachlan J, Pasquinelli AE, Bálint E, Tuschl T, Zamore PD (2001) A cellular function for the RNA-interference enzyme Dicer in the maturation of the let-7 small temporal RNA. Science 293(5531):834–838
34. Liu J, Carmell MA, Rivas FV, Marsden CG, Thomson JM, Song JJ, Hammond SM, Joshua-Tor L, Hannon GJ (2004) Argonaute2 is the catalytic engine of mammalian RNAi. Science 305(5689):1437–1441
35. Winter J, Jung S, Keller S, Gregory RI, Diederichs S (2009) Many roads to maturity: microRNA biogenesis pathways and their regulation. Nat Cell Biol 11(3):228–234
36. Jackson A, Linsley PS (2010) The therapeutic potential of microRNA modulation. Discov Med 9(47):311–318
37. Jackson AL, Linsley PS (2010) Recognizing and avoiding siRNA off-target effects for target identification and therapeutic application. Nat Rev Drug Discov 9(1):57–67
38. Astashkina A, Mann B, Grainger DW (2012) A critical evaluation of in vitro cell culture models for high-throughput drug screening and toxicity. Pharmacol Ther 134(1):82–106
39. Lipinski C, Hopkins A (2004) Navigating chemical space for biology and medicine. Nature 432(7019):855–861
40. Gumireddy K, Young D, Xiong X, Hogenesch J, Huang Q, Deiters A (2008) Small-molecule inhibitors of microRNA miR-21 function. Angew Chem Int Ed 47(39):7482–7484
41. Young D, Connelly C, Grohmann C, Deiters A (2010) Small molecule modifiers of MicroRNA miR-122 function for the treatment of hepatitis C virus infection and hepatocellular carcinoma. J Am Chem Soc 132(23):7976–7981
42. Chandrasekhar S, Pushpavalli SN, Chatla S, Mukhopadhyay D, Ganganna B, Vijeender K, Srihari P, Reddy CR, Janaki Ramaiah M, Bhadra U (2012) aza-Flavanones as potent cross-species microRNA inhibitors that arrest cell cycle. Bioorg Med Chem Lett 22(1): 645–648
43. Michlewski G, Guil S, Semple CA, Cáceres JF (2008) Posttranscriptional regulation of miRNAs harboring conserved terminal loops. Mol Cell 32(3):383–393
44. Guil S, Cáceres JF (2007) The multifunctional RNA-binding protein hnRNP A1 is required for processing of miR-18a. Nat Struct Mol Biol 14(7):591–596
45. Lünse CE, Michlewski G, Hopp CS, Rentmeister A, Cáceres JF, Famulok M, Mayer G (2010) An aptamer targeting the apical-loop domain modulates pri-miRNA processing. Angew Chem Int Ed Engl 49(27):4674–4677
46. Ellington AD (1994) RNA selection. Aptamers achieve the desired recognition. Curr Biol 4(5):427–429
47. Krishnamurthy M, Simon K, Orendt AM, Beal PA (2007) Macrocyclic helix-threading peptides for targeting RNA. Angew Chem Int Ed Engl 46(37):7044–7047
48. Lee Y, Hyun S, Kim HJ, Yu J (2008) Amphiphilic helical peptides containing two acridine moieties display picomolar affinity toward HIV-1 RRE and TAR. Angew Chem Int Ed Engl 47(1):134–137
49. Chirayil S, Chirayil R, Luebke KJ (2009) Discovering ligands for a microRNA precursor with peptoid microarrays. Nucleic Acids Res 37(16):5486–5497
50. Kwon YU, Kodadek T (2007) Quantitative evaluation of the relative cell permeability of peptoids and peptides. J Am Chem Soc 129(6): 1508–1509

51. Bose D, Jayaraj G, Suryawanshi H, Agarwala P, Pore SK, Banerjee R, Maiti S (2012) The tuberculosis drug streptomycin as a potential cancer therapeutic: inhibition of miR-21 function by directly targeting its precursor. Angew Chem Int Ed Engl 51(4):1019–1023
52. Kloosterman WP, Lagendijk AK, Ketting RF, Moulton JD, Plasterk RH (2007) Targeted inhibition of miRNA maturation with morpholinos reveals a role for miR-375 in pancreatic islet development. PLoS Biol 5(8):e203
53. Davies BP, Arenz C (2006) A homogenous assay for micro RNA maturation. Angew Chem Int Ed Engl 45(33):5550–5552
54. Davies BP, Arenz C (2008) A fluorescence probe for assaying micro RNA maturation. Bioorg Med Chem 16(1):49–55
55. Klemm C, Berthelmann A, Neubacher S, Arenz C (2009) Short and efficient synthesis of alkyne-modified amino glycoside building blocks. Eur J Org Chem 17:2788–2794
56. Neubacher S, Dojahn CM, Arenz C (2011) A rapid assay for miRNA maturation by using Unmodified pre-miRNA. Chembiochem 12(15):2302–2305
57. Maiti M, Nauwelaerts K, Herdewijn P (2012) Pre-microRNA binding aminoglycosides and antitumor drugs as inhibitors of Dicer catalyzed microRNA processing. Bioorg Med Chem Lett 22(4):1709–1711
58. Henn A, Joachimi A, Gonçalves DP, Monchaud D, Teulade-Fichou MP, Sanders JK, Hartig JS (2008) Inhibition of dicing of guanosine-rich shRNAs by quadruplex-binding compounds. Chembiochem 9(16):2722–2729
59. Shan G, Li Y, Zhang J, Li W, Szulwach KE, Duan R, Faghihi MA, Khalil AM, Lu L, Paroo Z, Chan AW, Shi Z, Liu Q, Wahlestedt C, He C, Jin P (2008) A small molecule enhances RNA interference and promotes microRNA processing. Nat Biotechnol 26(8):933–940
60. Zhang Q, Zhang C, Xi Z (2008) Enhancement of RNAi by a small molecule antibiotic enoxacin. Cell Res 18(10):1077–1079
61. Watashi K, Yeung ML, Starost MF, Hosmane RS, Jeang KT (2010) Identification of small molecules that suppress microRNA function and reverse tumorigenesis. J Biol Chem 285(32):24707–24716
62. Ebert MS, Neilson JR, Sharp PA (2007) MicroRNA sponges: competitive inhibitors of small RNAs in mammalian cells. Nat Methods 4(9):721–726
63. Zheng G, Cochella L, Liu J, Hobert O, Li WH (2011) Temporal and spatial regulation of microRNA activity with photoactivatable cantimirs. ACS Chem Biol 6(12):1332–1338
64. Achenbach JC, Chiuman W, Cruz RP, Li Y (2004) DNAzymes: from creation in vitro to application in vivo. Curr Pharm Biotechnol 5(4):321–336
65. Jadhav VM, Scaria V, Maiti S (2009) Antagomirzymes: oligonucleotide enzymes that specifically silence microRNA function. Angew Chem Int Ed Engl 48(14):2557–2560
66. An CI, Trinh VB, Yokobayashi Y (2006) Artificial control of gene expression in mammalian cells by modulating RNA interference through aptamer-small molecule interaction. RNA 12(5):710–716

Part II

Methods

Chapter 4

Expression Profiling of Components of the miRNA Maturation Machinery

Michael Sand and Marina Skrygan

Abstract

The quantitative real-time polymerase chain reaction (qRT-PCR) is a valuable and well-proven technique used to investigate the expression level of multiple components of the microRNA (miRNA) maturation machinery. Here, we describe how to determine the messenger RNA expression levels of components of the miRNA machinery starting from the isolation of the RNA from a tissue biopsy to performance of the qRT-PCR.

Key words MicroRNA maturation, Quantitative real-time polymerase chain reaction (qRT-PCR)

1 Introduction

Every protein has a characteristic amino acid sequence, which is determined by the coding gene. A gene is not immediately translated into a protein; it is expressed via the intermediate production of messenger RNA (mRNA). Chemically similar to DNA, RNA is a straight, high-molecular-weight polymer. mRNA is synthesized by the same process of complementary base pairing as occurs during the replication of DNA; however, mRNA is read from one strand of the DNA double helix, resulting in a single strand that is complementary only to that DNA strand. Apart from the substitution of thymine with uracil, the mRNA strand generated is identical to the remaining DNA strand. In addition, the mRNA contains a sequence of nucleotides that is composed of a sequence of exons, which corresponds to the sequence of amino acids in the protein. The intron regions of the DNA that do not contain amino acid coding information are excised during the synthesis of the mRNA. The transcription of a particular mRNA begins when a cell needs a particular protein.

Christoph Arenz (ed.), *miRNA Maturation: Methods and Protocols*, Methods in Molecular Biology, vol. 1095, DOI 10.1007/978-1-62703-703-7_4,

The microRNA (miRNA) maturation machinery is composed of a well-defined set of proteins including the intranuclear RNase III endonuclease Drosha and its cofactor DiGeorge syndrome critical region gene 8 (DGCR8 or Pasha), the miRNA nuclear export receptor RAN GTPase exportin-5 (Exp5), and the RNase III enzyme Dicer and its cofactors the double-stranded RNA-binding domain (dsRBD) protein Tar RNA-binding protein (TRBP) and the protein activator of PKR (PACT). Recently, a number of studies have provided growing evidence for a role of the miRNA maturation machinery in a variety of diseases, suggesting that it could serve as a prognostic tool [1–10]. In particular, the miRNA maturation machinery has been investigated in a variety of different tumors.

In this chapter, we share our experience of the expression profiling of components of the miRNA maturation machinery in the skin, which can be analogously performed with other tissue [11, 12].

2 Materials

1. RNA*later*® RNA Stabilization Reagent (QIAGEN).
2. RNeasy® Lipid Tissue Mini Kit (QIAGEN).
3. Stainless steel beads (5 mm).
4. Chloroform (99 %, molecular biology grade).
5. Isopropanol (99.7 % analytical grade).
6. DNAse I (RNAse free, 1,500 U).
7. DNase I (10 U/μl; RNase free).
8. TaqMan® Reverse Transcription Reagents Kit (Applied Biosystems).
9. *Power*SYBR Green PCR Mix (Applied Biosystems).
10. Gel-loading buffer.
11. Agarose HR-Plus.
12. NuSieve agarose.
13. 50× TAE buffer (commercial).
14. 100-bp DNA ladder.
15. 1 % ethidium bromide solution.

3 Methods

3.1 Isolation of Total RNA from a Tissue Biopsy

RNAlater-fixed tissue can be stored for 1 day at 37 °C, for 7 days at 15–25 °C, or 4 weeks at 2–8 °C (*see* **Notes 1** and **2**). For longer storage periods, the fixed tissue should be maintained at −20 or −80 °C. Total cellular RNA is isolated using the RNeasy® Lipid

Tissue Mini Kit in accordance with the manufacturer's protocol (*see* **Note 3**).

1. The tissue is transferred from the tube containing the RNAlater solution to a fresh 2.0-ml Eppendorf tube containing 1 ml of the QIAzol® lysis reagent and a stainless steel bead (*see* **Note 4**). The tissue is subjected to shredding at 50 Hz for 10 min at 15–25 °C using a TissueLyser LT (QIAGEN) (*see* **Note 5**).
2. The stainless steel bead is removed, and 200 μl chloroform is added to the tissue lysate. The mixture is vortexed thoroughly and centrifuged at maximum speed for 15 min at 4 °C.
3. After centrifugation, the mixture of tissue lysate and chloroform has separated into three phases. The upper aqueous phase contains the RNA, the interphase contains the DNA, and the lower phase contains the chloroform and proteins. The upper RNA-containing phase is transferred into a fresh 1.5-ml Eppendorf tube, 600 μl isopropanol is added, and the mixture is vortexed thoroughly for 1 min (*see* **Note 6**).
4. A 700-μl sample of the solution is transferred to an RNeasy® Mini Spin column and centrifuged at 10,000 rpm for 15 s. The flow-through is discarded, and the RNA is bound to the quartz membrane of the RNeasy column.
5. **Step 4** is repeated using the remainder of the RNA-containing mixture.
6. To remove any membrane-bound chromosomal DNA, the membrane is treated with DNAse I. The DNAse I solution is prepared by mixing 70 μl RDD (RNeasy®) buffer with 10 μl DNase I. This solution (80 μl) is pipetted onto the middle of the column, and the column is incubated for 19 min at 15–25 °C.
7. The column is washed with 700 μl RW1 buffer (RNeasy®) under the same centrifugation conditions as previously described. The flow-through is discarded.
8. The column is washed twice with 500 μl RPE buffer containing ethanol (RNeasy®). The column is centrifuged using the conditions described above.
9. The column is dried by centrifugation for 2 min at maximum speed.
10. The column is transferred into a 1.5-ml Eppendorf tube.
11. To elute the total RNA from the column, 30 μl RNase-free water is pipetted onto the quartz membrane.
12. After incubation for 1 min at 15–25 °C, followed by centrifugation for 1 min at 11,000 rpm, the eluted RNA solution is frozen and stored at −80 °C.

3.2 Reverse Transcribing mRNA into cDNA

Using reverse transcription, mRNA produces a single-stranded complementary DNA (cDNA) sequence. The cDNA is not present in the cell under physiologic conditions, is an accurate reflection of the mRNA, and is easier to examine. To prepare a cDNA, a pd(N)6 primer is bound to one end of the mRNA molecule. This double-stranded area is the starting point for the reverse transcriptase enzyme, which can produce a single-stranded DNA copy. The cDNA chain is extended in the 5′–3′ direction, and the deoxyribonucleotides are added using the RNA sequence as a template. The result is a hybrid molecule consisting of the mRNA template strand and the complementary DNA strand, which is bound through Watson-and-Crick base pairing. The single-stranded cDNA can be separated from the RNA strand using heat. This cDNA can be amplified further using the PCR, which produces a large quantity of the DNA sequence that represents the mRNA.

1. To remove any genomic DNA residues, the total RNA is treated with DNase I for a second time before it is rewritten into cDNA (*see* **Note 7**). The DNase I reaction is performed as follows: 10 μl RNA solution, 1.5 μl 10× RT buffer, 3.3 μl MgCl2 (25 mM), 0.5 μl RNase inhibitor (20 U/μl), and 0.3 μl DNase I are placed into 0.5-ml Eppendorf tubes, which are vortexed thoroughly.
2. The mixture is incubated in a thermoblock (Biometra, Germany) for 20 min at 37 °C, followed by 8 min at 75 °C, and the mixture is cooled briefly on ice.
3. The total RNA is transcribed into cDNA using the TaqMan® Reverse Transcription Reagent Kit (Applied Biosystems, USA) as follows: 15 μl of the DNase I-treated RNA solution is added to 1.5 μl 10× RT buffer, 3.3 μl $MgCl_2$ (25 mM), 6.0 μl dNTPs, 1.5 μl random hexamer primers (5 nM), 0.5 μl RNase inhibitor (20 U/μl), and 0.8 μl MultiScribe™ reverse transcriptase (50 U/μl), mixed thoroughly, and diluted to a final volume of 30 μl using nuclease-free water.
4. The sample is incubated for 10 min at 25 °C, 45 min at 48 °C, and 5 min at 95 °C using a Trio-Thermoblock (Biometra).
5. The prepared cDNA solution is stored at −20 °C until further processing.

3.3 Real-Time Polymerase Chain Reaction

Developed by Mullis and Fallona in 1987, the polymerase chain reaction (PCR) can selectively multiply desired target genome segments. PCR uses DNA polymerases that can polymerize a single DNA strand into a double strand provided that a short double-stranded region is available as a primer. The first step is to denature the double-stranded DNA into two single strands by heating to 95 °C, which enables the binding of the primer. Next, the single strands are hybridized using two sequence-specific oligonucleotides (primers), which are complementary to the sequences on either side of the target region. This step is termed the "annealing" step and

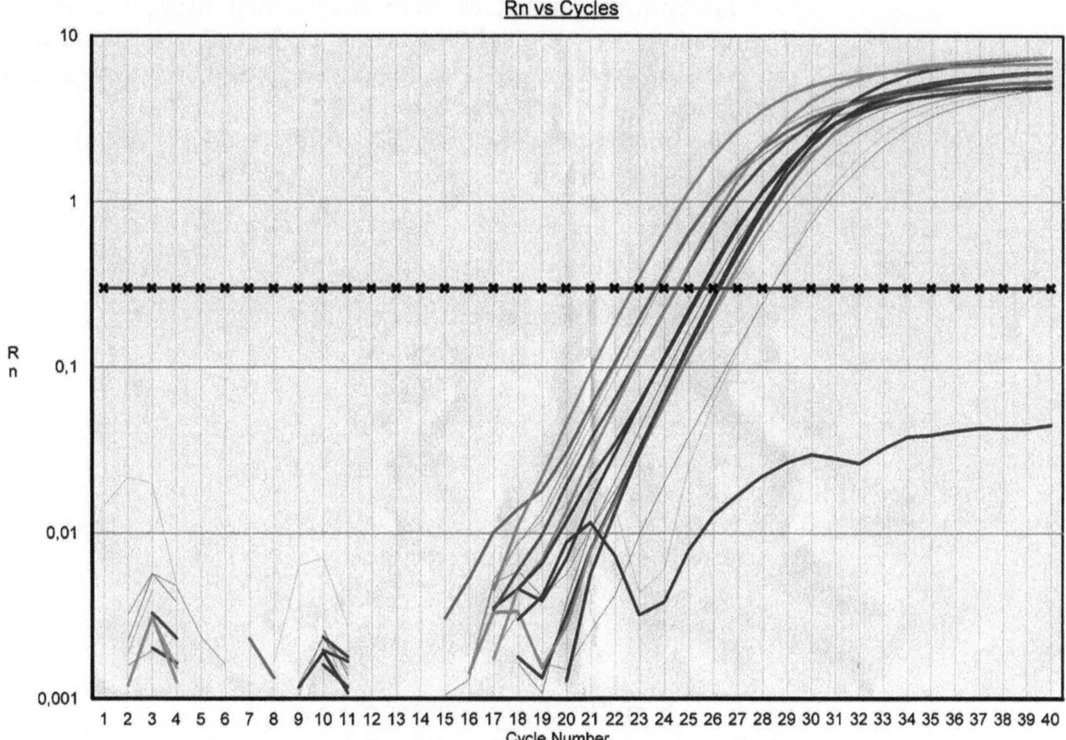

Fig. 1 Fluorescence curves for PCR amplifications

takes place during the cooling down period of the reaction. The "annealing temperature" depends on the GC content of the primers; the higher the GC content, the more specifically the oligonucleotides bind and the less the nonspecific products generated. In the extension step, the DNA polymerase synthesizes complementary DNA fragments starting at the 3′-OH ends of the two primers. The temperature used is optimal for the DNA polymerase. The resulting double-stranded DNA sequence is denatured again, and the process is repeated from the beginning of the annealing step. For every round of PCR, the desired sequence is doubled.

Real-time PCR was developed by Higuchi et al. in 1993 for the quantitative determination of the transcript levels of individual target genes. The main principle of this method is that the PCR preparation is expanded by adding a fluorescent dye (e.g., SYBR® Green I), which intercalates between the newly synthesized DNA double strands during amplification. The resulting increase in fluorescence is monitored and measured during the PCR reaction.

To quantify the transcript, a straight line is drawn through the linear range of the amplification curve (threshold), and the cycle number (Ct value) is determined.

The stronger the target gene is expressed, the sooner a linear gradient and smaller Ct values are observed on the fluorescence curve (Fig. 1).

The primer sequences were determined using the Primer Express software (Applied Biosystems, USA) and were synthesized by a custom service provider that specializes in the synthesis of oligonucleotides (*see* **Notes 8** and **9**).

The primers that we have successfully used for amplification include the following:

Dicer1

Forward 5′-ttaacctttggtgtttgatgagtgt-3′.

Reverse 5′-ggacatgatggacaattttcaca-3′.

Drosha

Forward 5′-catgtcacagaatgtcgttcca-3′.

Reverse 5′-gggtgaagcagcctcagattt-3′.

PACT

Forward 5′-tgcagttcctgacccccttaatg-3′.

Reverse 5′-agccaattcctgtaatgaaccaa-3′.

DGCR

Forward 5′-gcaagatgcacccacaaaga-3′.

Reverse 5′-ttgaggacacgctgcatgtac-3′.

TARBP1

Forward 5′-cattaatggatgcgctttcaga-3′.

Reverse 5′-tgtaatttcagtcccaatggagaac-3′.

TARBP2

Forward 5′-ggttgccggagtacacagtga-3′.

Reverse 5′-tgccactcccaatctcaatg-3′.

1. PCR mixture:
   - 5 μl cDNA solution.
   - 12.5 μl 2× *Power*SYBR Green PCR Mix (Applied Biosystems).
   - 5 μM of each primer and diethylpyrocarbonate (DEPC)-treated water to a final volume of 25 μl.
2. The amplification reaction is performed in the StepOne Real-Time PCR System (Applied Biosystems, USA) using the following temperature protocol:
   (a) For PCR master mix activation: 10 min at 95 °C.
   (b) PCR: 15 s at 95 °C and 60 s at 60 °C.

   Because the SYBR® Green dye binds both the specific and nonspecific DNA, the double-stranded PCR products are melted at 95 °C after 40 cycles. This process linearly decreases

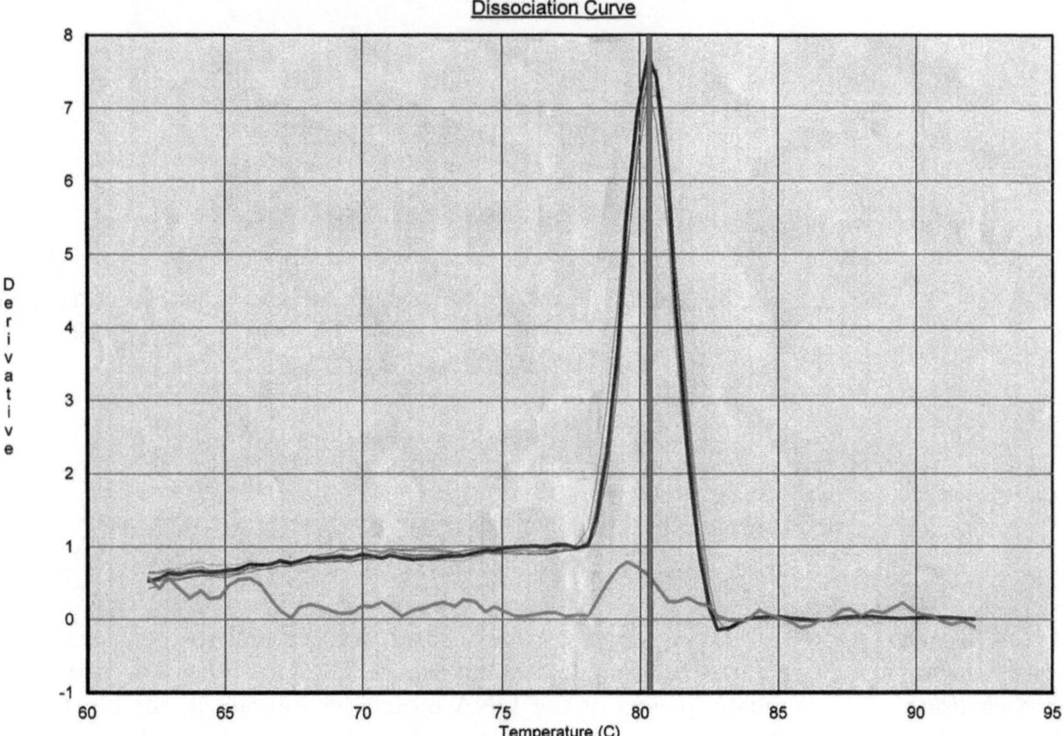

Fig. 2 Dissociation (melting) curve of the PCR products

the fluorescence intensity of the dye until it reaches the melting temperature, which is specific for the DNA target sequence. For example, the melting temperature for Dicerl is 76 °C. When the melting temperature is reached, the DNA fragment denatures into a single strand and releases the entire fluorescent dye. The specific melting temperature depends on the size of the PCR amplificate; therefore, the nonspecific products and primer dimers melt at different temperatures (*see* **Note 10**). A melting or a dissociation curve specifically detects the amplified target products (Fig. 2). In addition, the amplificate size is monitored using agarose gel electrophoresis.

3. The housekeeping genes (HKGs) GAPDH and RPL38 were investigated for their expression stability. The expression level of RPL38 was shown to be the least regulated of all the HKGs studied; therefore, RPL38 was used as the reference gene for the relative quantification of the PCR products (*see* **Note 11**).

RPL 38

Forward 5′-TCACTGACAAAGAGAAGGCAGAGA-3′.

Reverse 5′-TCAGTGTGTCTGGTTCATTTCAGTT-3′.

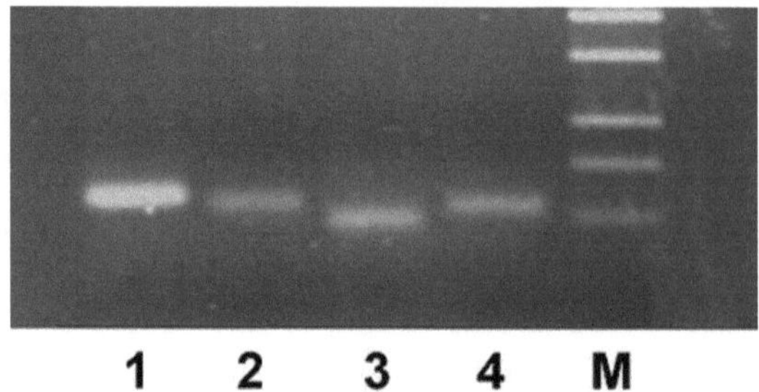

Fig. 3 Agarose gel showing DNA fragments specific to our PCR products ((1) Drosha, 115 bp; (2) EIF2C1, 105 bp; (3) DGCR8, 93 bp; (4) TARBP2, 102 bp); the *right column* (M) shows the 100-bp DNA size standard

GAPDH

Forward 5′-TCACTGACAAAGAGAAGGCAGAGA-3′.

Reverse 5′-TCAGTGTGTCTGGTTCATTTCAGTT-3′.

3.4 Verification of the Amplification Products Using Agarose Gel Electrophoresis

DNA fragments can be separated according to their size using agarose gels and an electric field. The negatively charged phosphate groups of the DNA molecules migrate in the electrically neutral agarose gel matrix toward the anode, and the smaller fragments move more rapidly than the large fragments. The accuracy of the PCR is evaluated by the size of the DNA fragments:

1. 5 μl of the PCR products is mixed with 5 μl of gel loading buffer.
2. A 100-bp DNA size standard (these DNA standards contain DNA fragments of precisely defined sizes at intervals of 100 bp) is applied to the 2 % agarose gel (1 % agarose, 1 % NuSieve agarose, and 1× TAE buffer) (*see* **Note 12**).
3. The electrophoresis is performed for 1 h at 175 V and 0.08 A. The agarose gel is removed from the chamber and transferred into a bowl containing the ethidium bromide solution for 15 min.
4. The DNA fragments are visualized under UV light and are recorded (Fig. 3).

3.5 Calculating the Results of the PCR Using Relative Quantification

For the relative quantitation of the PCR, the expression of the target gene is correlated to a non-regulated homogeneously expressed HKG or to an HKG index composed of several HKGs. Different RNA extraction efficiencies and errors in the reverse transcriptase reaction are expected for every sample of both the target gene and the HKG. Therefore, when calculating the differences in the expression of these genes, the individual sample effects cancel each other out.

The relative expression of the target gene is compared to the control. The difference in the expression is calculated as follows:

$\Delta Ct = Ct$ target gene − Ct HKG.

Ratio = 2 − ΔCt.

This calculation method assumes that the amount of DNA in each cycle doubles; therefore, the real-time PCR efficiency is optimal.

4 Notes

1. For quantitative mRNA expression analysis, the tissue specimens should be fixed until processed in RNAlater® Stabilization Reagent. This step is necessary to prevent the denaturation of the mRNA by ribonucleases (RNases). RNases are stable enzymes that exhibit high endonuclease activity that specifically cleaves RNAs. These enzymes are produced by all living organisms and are found in practically every location.
2. After harvesting, the tissue specimen should be transferred immediately to RNAlater stabilization reagent for storage. The ratio of tissue to fixative solution should be at least 1:6, although a 1:10 ratio is ideal.
3. Gloves should be worn at all times when handling RNAs to prevent the contamination of the specimen with RNases.
4. Without exception, consumables (pipette tips, Eppendorf tubes, etc.) should be RNase free.
5. Prior to the chemical treatment, the fixed tissue should first be crushed mechanically.
6. For the precipitation of nucleic acids, isopropanol is more efficient than 70 % ethanol.
7. For the second round of DNase I treatments, we recommend using a different manufacturer because enzymes from different companies often differ greatly in their quality and effects.
8. For successful amplification, the primers should meet the following criteria:
   - Primer length: 16–30 bp; 20–25 bp is optimal.
   - GC content: 50–60 %.
   - Primer Tm: 58–60 °C; the maximum Tm difference between the two primers should not exceed 2 °C.
9. The primers should be tested for mRNA specificity using the BLAST method (http://blast.ncbi.nlm.nih.gov/Blast.cgi) to prevent nonspecific primer binding and therefore nonspecific gene amplification.

10. For an optimal real-time PCR efficiency of 2.00 per cycle, the amplicon length should not exceed 150 bp.
11. When selecting the appropriate HKG, the following guidelines should be considered:

    The expression level of the HKG should be approximately the same Ct value, whereas the Ct values of the target genes can differ. Ideally, the change in the mean (the sample mean minus the control mean) for HKGs should be between 0.5 and 0.9 and should not exceed 1. If the change in the mean is greater than 1.5, it cannot be accepted as "not regulated"; therefore, it cannot be used as an appropriate HKG for further calculations.
12. For the separation of shorter fragments, a mixture of HR-agarose and NuSieve agarose is recommended. BMA NuSieve 3:1 agarose is suitable for the fine separation of small (less than 1,000 bp) DNA fragments because of its high melting point (<90 °C) and low background fluorescence. The NuSieve agarose alone is very brittle and is difficult to handle; therefore, it is recommended to mix the NuSieve with normal agarose at a ratio of 1:1.

References

1. Karube Y, Tanaka H, Osada H, Tomida S, Tatematsu Y, Yanagisawa K, Yatabe Y, Takamizawa J, Miyoshi S, Mitsudomi T, Takahashi T (2005) Reduced expression of Dicer associated with poor prognosis in lung cancer patients. Cancer Sci 96(2):111–115. doi:CAS015 [pii]10.1111/j.1349-7006.2005.00015.x
2. Chiosea SI, Barnes EL, Lai SY, Egloff AM, Sargent RL, Hunt JL, Seethala RR (2008) Mucoepidermoid carcinoma of upper aerodigestive tract: clinicopathologic study of 78 cases with immunohistochemical analysis of Dicer expression. Virchows Arch 452(6):629–635. doi:10.1007/s00428-007-0574-5
3. Grelier G, Voirin N, Ay AS, Cox DG, Chabaud S, Treilleux I, Leon-Goddard S, Rimokh R, Mikaelian I, Venoux C, Puisieux A, Lasset C, Moyret-Lalle C (2009) Prognostic value of Dicer expression in human breast cancers and association with the mesenchymal phenotype. Br J Cancer 101(4):673–683. doi:6605193 [pii]10.1038/sj.bjc.6605193
4. Faggad A, Budczies J, Tchernitsa O, Darb-Esfahani S, Sehouli J, Muller BM, Wirtz R, Chekerov R, Weichert W, Sinn B, Mucha C, Elwali NE, Schafer R, Dietel M, Denkert C (2010) Prognostic significance of Dicer expression in ovarian cancer-link to global microRNA changes and oestrogen receptor expression. J Pathol 220(3):382–391. doi:10.1002/path.2658
5. Dedes KJ, Natrajan R, Lambros MB, Geyer FC, Lopez-Garcia MA, Savage K, Jones RL, Reis-Filho JS (2011) Down-regulation of the miRNA master regulators Drosha and Dicer is associated with specific subgroups of breast cancer. Eur J Cancer 47(1):138–150. doi:S0959-8049(10)00788-4 [pii]10.1016/j.ejca.2010.08.007
6. Valencak J, Schmid K, Trautinger F, Wallnofer W, Muellauer L, Soleiman A, Knobler R, Haitel A, Pehamberger H, Raderer M (2011) High expression of Dicer reveals a negative prognostic influence in certain subtypes of primary cutaneous T cell lymphomas. J Dermatol Sci 64(3):185–190. doi:S0923-1811(11)00259-3 [pii]10.1016/j.jdermsci.2011.08.011
7. Guo X, Liao Q, Chen P, Li X, Xiong W, Ma J, Luo Z, Tang H, Deng M, Zheng Y, Wang R, Zhang W, Li G (2012) The microRNA-processing enzymes: Drosha and Dicer can predict prognosis of nasopharyngeal carcinoma. J Cancer Res Clin Oncol 138(1):49–56. doi:10.1007/s00432-011-1058-1
8. Papachristou DJ, Rao UN, Korpetinou A, Giannopoulou E, Sklirou E, Kontogeorgakos V, Kalofonos HP (2012) Prognostic significance of Dicer cellular levels in soft tissue sarcomas. Cancer Invest 30(2):172–179. doi:10.3109/07357907.2011.633293
9. Zhu DX, Fan L, Lu RN, Fang C, Shen WY, Zou ZJ, Wang YH, Zhu HY, Miao KR, Liu P,

Xu W, Li JY (2012) Downregulated Dicer expression predicts poor prognosis in chronic lymphocytic leukemia. Cancer Sci 103(5):875–881. doi:10.1111/j.1349-7006.2012.02234.x

10. Shu GS, Yang ZL, Liu DC (2012) Immunohistochemical study of Dicer and Drosha expression in the benign and malignant lesions of gallbladder and their clinicopathological significances. Pathol Res Pract 208(7):392–397. doi:S0344-0338(12)00125-2 [pii]10.1016/j.prp.2012.05.001
11. Sand M, Gambichler T, Skrygan M, Sand D, Scola N, Altmeyer P, Bechara FG (2010) Expression levels of the microRNA processing enzymes Drosha and dicer in epithelial skin cancer. Cancer Invest 28(6):649–653. doi:10.3109/07357901003630918
12. Sand M, Skrygan M, Georgas D, Arenz C, Gambichler T, Sand D, Altmeyer P, Bechara FG (2011) Expression levels of the microRNA maturing microprocessor complex component DGCR8 and the RNA-induced silencing complex (RISC) components Argonaute-1, Argonaute-2, PACT, TARBP1, and TARBP2 in epithelial skin cancer. Mol Carcinog. doi:10.1002/mc.20861

Chapter 5

Primary MicroRNA Processing Assay Reconstituted Using Recombinant Drosha and DGCR8

Ian Barr and Feng Guo

Abstract

In animals, the Microprocessor complex cleaves primary transcripts of microRNAs (pri-miRNAs) to produce precursor microRNAs in the nucleus. The core components of Microprocessor include the Drosha ribonuclease and its RNA-binding partner protein DiGeorge critical region 8 (DGCR8). DGCR8 has been shown to tightly bind an Fe(III) heme cofactor, which activates its pri-miRNA processing activity. Here we describe how to reconstitute pri-miRNA processing using recombinant human Drosha and DGCR8 proteins. In particular, we present the procedures for expressing and purifying DGCR8 as an Fe(III) heme-bound dimer, the most active form of this protein, and for estimating its heme content.

Key words RNA processing, DiGeorge syndrome, Heme, RNA-binding protein, Nucleic acid-binding protein, Pasha, Ribonuclease III

1 Introduction

1.1 The Microprocessor Complex

The Microprocessor complex minimally contains the proteins Drosha [1], an RNase III family member, and DGCR8 [2, 3], an RNA-binding protein that also contains the cofactor heme [4–6]. Drosha and DGCR8 are essential for processing of all canonical microRNAs (miRNAs) in animals [7–9]. They are also sufficient to reconstitute pri-miRNA processing activity in vitro [3, 10]. Furthermore, quite a few proteins have been shown to regulate the Drosha/DGCR8-mediated cleavage of pri-miRNAs [11–18].

The domain structures of Drosha and DGCR8 have been dissected in several studies. Drosha contains two RNase III domains and a double-stranded RNA-binding domain (dsRBD) in the C-terminal region (Fig. 1a). The central region of Drosha, including residues 390–900, is highly conserved and required for pri-miRNA processing but contains no recognizable sequence motifs [10, 19]. DGCR8 contains a heme-binding domain, two dsRBDs, and a C-terminal tail (CTT) (Fig. 1b). The dsRBDs contribute to pri-miRNA binding [4, 20, 21]. The C-terminal tail has been

Christoph Arenz (ed.), *miRNA Maturation: Methods and Protocols*, Methods in Molecular Biology, vol. 1095,
DOI 10.1007/978-1-62703-703-7_5, © Springer Science+Business Media New York 2014

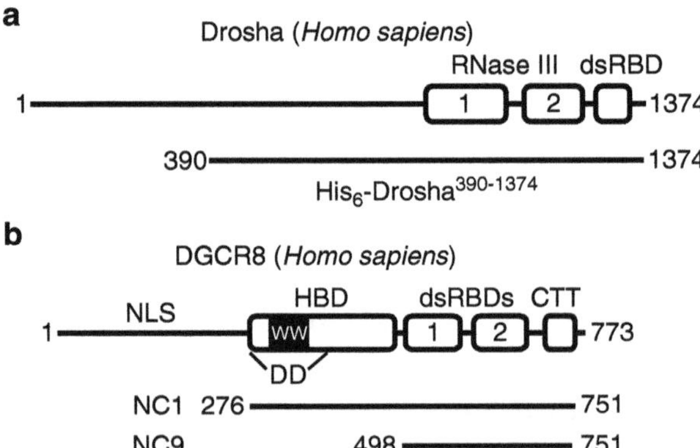

Fig. 1 Domain structures and recombinant expression constructs of human Drosha and DGCR8. The heme-binding domain (HBD) of DGCR8 includes a dimerization (sub)domain (DD)

shown to be important for co-immunoprecipitation of DGCR8 with Drosha in human cells [10] and for formation of proper higher order structure of DGCR8 upon binding pri-miRNAs [22]. See a recent review for more background information [23].

pri-miRNA processing may be analyzed in vitro using either whole-cell or nuclear extracts [24, 25], affinity-purified Microprocessor complexes expressed in mammalian cells [19, 26–28], or recombinant Drosha and DGCR8 expressed in heterologous systems [3, 10]. We focus on the last method in this review. Insect cell and bacterial expression systems allow active Drosha and DGCR8 proteins to be expressed with high yield and be purified to near homogeneity. These highly purified proteins do not contain other human proteins typically found to associate with Drosha and DGCR8 and enable the investigation of pri-miRNA processing with greater control of the experimental conditions.

Gregory, Shiekhattar, and colleagues were the first to show that Microprocessor may be reconstituted by expressing Drosha in insect cells and DGCR8 in *E. coli* [3]. The N-terminal 275 amino acids [10, 20] and the C-terminal 22/23 residues [4, 20] of the 773-residue DGCR8 have been shown to be dispensable for in vitro pri-miRNA processing. A DGCR8 construct with these residues deleted (named NC1, Fig. 1b) is the most active form in pri-miRNA processing in vitro, whereas a further truncation called NC9 (Fig. 1b), containing only the two dsRBDs and the CTT, is less active than NC1 [6]. The procedures for expression and purification of NC1 are described in this review.

1.2 DGCR8 as a Heme Protein

In addition to being an obligate partner of Drosha, DGCR8 also binds heme [4]. The central region of DGCR8, including the WW motif, encodes a unique dimeric heme-binding domain [29]. Each DGCR8 dimer binds one heme molecule [4]. The heme in native DGCR8, expressed in *E. coli*, is in the Fe(III) redox state [5].

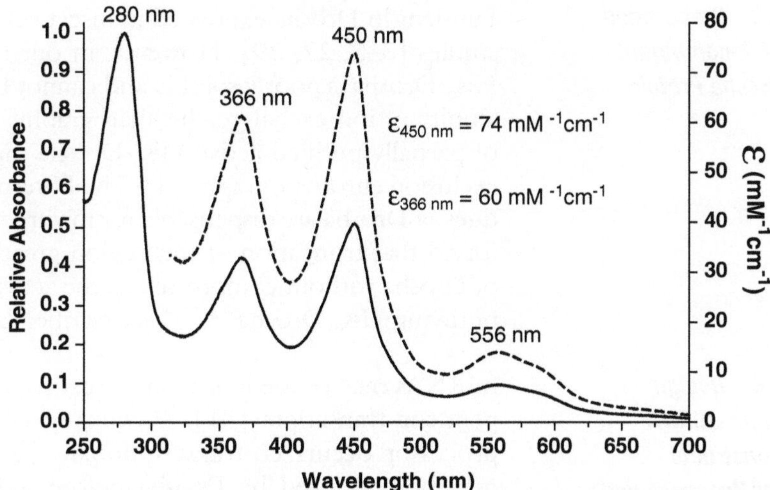

Fig. 2 Electronic absorbance spectrum of Fe(III) heme-bound NC1 dimer. The *solid line*, corresponding to the *left y*-axis, shows the relative absorbencies of the heme and protein peaks. The *dashed line*, corresponding to the *right y*-axis, shows the extinction coefficients of the heme, as determined using the pyridine hemochromagen assay as recently reported [30]

DGCR8 ligates to the Fe(III) using two Cys352 side chains contributed by both subunits; this coordination configuration results in characteristic absorption peaks at 366, 450, and 556 nm (Fig. 2) [5]. Recently, we show that Fe(III) heme activates dimeric apoNC1 for pri-miRNA processing in vitro, whereas Fe(II) heme does not [6]. Dimerization and heme binding are likely conserved properties of DGCR8 in all vertebrates and at least some invertebrates such as the star fish *Patiria miniata* [30]. Heme and the heme-binding domain appear to be important for pri-miRNA processing both in vitro and in vivo ([4, 6, 29] and our unpublished data), though their physiological functions have not been determined.

In order to obtain consistent results in studying pri-miRNA processing and to interpret them properly, it is important to express and purify recombinant DGCR8 protein with optimal heme content. When NC1 is overexpressed in *E. coli*, a heme-deficient condition is generated and hence some heme-free protein is produced [4]. The heme-free NC1 may appear as dimer and monomer [4, 6]. At least a part of the latter species is actually heterodimer of NC1, in which a subunit is cleaved by bacterial proteases during overexpression and/or purification so that only a small fragment (the dimerization domain) is left bound to the intact subunit [29]. The heme content of DGCR8 is indicated by the $A_{450\,nm}/A_{280\,nm}$ ratio, if the protein is purified to be free of nucleic acids from expression hosts. It is also possible to prepare apoNC1 from Fe(III) heme-bound NC1 via reduction and heme removal for studying activation of DGCR8 by heme [6]. However, the preparation of apoNC1 is beyond the scope of this chapter.

1.3 Preparation of Recombinant Drosha Protein

Full-length Drosha expressed in insect cells has been used in several studies [3–5, 22, 29]. However, in our experience the full-length His_6-Drosha is poorly soluble and cannot be purified using either Ni affinity or ion-exchange chromatography; and only a small amount of partially purified active His_6-Drosha may be obtained using size exclusion chromatography [4]. The N-terminal 390 amino acid residues of Drosha are dispensable for in vitro activity [10]. Recently, we found that truncation of this region greatly improves the solubility of Drosha without compromising the activity [6]. The procedure for purifying His_6-$Drosha^{390-1374}$ is described below.

1.4 Design of pri-miRNA Constructs for Reconstituted pri-miRNA Processing Assay

miRNAs may reside in introns or exons, messenger RNAs or independent transcripts [31]. Processing of pri-miRNAs by Microprocessor occurs co-transcriptionally [32]. Intronic pri-miRNAs may be processed by Drosha before splicing catalysis [33]. The exact 5′ and 3′ ends of pri-miRNAs at the time of processing are often not known. Biochemical studies show that pri-miRNA fragments containing the precursor miRNA (pre-miRNA) and certain lengths of the immediate flanking regions can be processed by Drosha and DGCR8 [1, 24]. The minimal lengths of the flanking regions for efficient processing by affinity-purified Microprocessor complexes may be as short as 10–20 nt [10, 19]. For reconstituted pri-miRNA processing assays, we typically include 30–60 nt on both sides of the pre-miRNA region.

The pri-miRNA fragments are typically prepared using T7 or SP6 RNA polymerase. The protocol for how to use T7 RNA polymerase is provided here. The transcription template should contain the T7 promoter, followed by the pri-miRNA coding sequence. For high transcription yields, the first two nucleotides of the transcript should be guanosines [34]. Either a PCR product or a linearized plasmid may serve as the template for the run-off transcription, in which the 3′-end of the RNA is roughly defined by the end of the template where the RNA polymerase simply falls off. The T7 RNA polymerase is known to add 0–3 non-templated residues at the 3′-end of the transcripts [34]. In the case where a plasmid template is used, a cleavage site for a restriction endonuclease such as *PstI* or *EarI* is engineered for linearization.

2 Materials

All solutions should be, to the greatest extent possible, free from RNase contamination. This applies especially to the reagents and buffers involved in transcription and pri-miRNA processing reactions and in the storage of RNAs.

2.1 Expression and Purification of DGCR8

1. The NC1 expression plasmid contains the coding sequence of amino acid residues 276–751 (NCBI accession no. of full-length DGCR8 cDNA: BC037564) inserted between *NdeI*

and *EcoRI* sites of pET-24a(+) (kanamycin resistant) or pET-17b (ampicillin resistant) vector. The PCR primers used in cloning are CAGCCATATGGATGGAGAGACAAGTGTGC (forward, the *NdeI* site underlined) and GCTCGAATTCAC TTTCGAGTCTCCTCCCT (reverse, the *EcoRI* site underlined).

2. *E. coli* strain BL21-CodonPlus (DE3)-RIPL (Agilent Technologies).
3. LB-Miller medium.
4. UV-visible absorption spectrophotometer equipped with a turbidity cuvette holder.
5. Isopropyl β-D-1-thiogalactopyranoside (IPTG).
6. δ-aminolevulinic acid (δ-ALA).
7. High-speed centrifuge.
8. Sonics Vibra-Cell VCX 750 ultrasonic processor equipped with a standard probe.
9. Chromatography systems such as ÄKTA Purifier and ÄKTA Prime.
10. 5-mL HiTrap SP HP cation exchange column (GE Healthcare).
11. Superdex 200 10/300 GL gel filtration column (GE Healthcare).
12. DGCR8 lysis buffer: 20 mM Tris–HCl pH 8.0, 100 mM NaCl, 1 mM phenylmethylsulfonyl fluoride (PMSF), 1 mM dithiothreitol (DTT) (*see* **Note 1**).
13. DGCR8 buffer *A*: 20 mM Tris–HCl pH 8.0, 100 mM NaCl, 1 mM DTT.
14. DGCR8 buffer *B*: 20 mM Tris–HCl pH 8.0, 2 M NaCl, 1 mM DTT.
15. SEC buffer: 20 mM Tris–HCl pH 8.0, 400 mM NaCl, 1 mM DTT.
16. Centrifugal concentrator with a molecular weight cutoff of 30 kDa.

2.2 Expression and Purification of Drosha

1. The His_6-Drosha$^{390-1374}$ expression plasmid has the coding sequence of amino acid residues 390–1,374 of human Drosha (NCBI accession no. of full-length cDNA: NM_013235) inserted between *BamHI* and *NotI* sites of pFastBac-HTb vector. The PCR primers used in cloning are CGCGGATCC AAAGAGCCCGAGGAGACC (forward, the *BamHI* site underlined) and GAGGATTAGAGCGGCCGCTTATTTCT TGATGTCTTCAGTCTC (reverse, the *NotI* site underlined). The recombinant His_6-Drosha$^{390-1374}$ contains a His_6-tag and a TEV cleavage site at its N-terminus (*see* **Note 2**).
2. Sf9 insect cells, culture medium, and transfection reagent.
3. Equipment same as the ones described for DGCR8 purification.

4. Ni Sepharose High Performance column (GE Healthcare).
5. Drosha lysis buffer: 20 mM Tris–HCl (pH 8.0), 500 mM NaCl, 10 mM imidazole, 20 % (v/v) glycerol, and 0.83 mM PMSF.
6. Drosha wash buffer: 20 mM Tris–HCl (pH 8.0), 500 mM NaCl, 10 mM imidazole, and 20 % (v/v) glycerol.
7. Drosha elution buffer: 20 mM Tris–HCl (pH 8.0), 500 mM NaCl, 200 mM imidazole, and 20 % (v/v) glycerol.
8. Drosha storage/reaction buffer: 20 mM Tris–HCl (pH 8.0), 100 mM NaCl, 0.2 mM ethylenediaminetetraacetic acid (EDTA), and 10 % (v/v) glycerol.
9. Liquid nitrogen or dry ice–ethanol mix.

2.3 Transcription and Purification of pri-miRNAs Uniformly Labeled with ^{32}P

1. The transcription template for a human pri-miR-30a fragment is a linearized pUC19 plasmid containing the sequence GAATTC*TAATACGACTCACTATA***GGAAAGAAGGTATATTGCTGTTGACAGTGAGCGACTGTAAACATCCTCGACTGGAAGCTGTGAAGCCACAGATGGGCTTTCAGTCGGATGTTTGCAGCTGCCTACTGCCTCGGACTTCAAGGGGCTACTTTAGGAGCAATTATCTTGTTT**CgaagagTCTAGA (the *EcoRI* and *XbaI* cloning sites are underlined; the T7 promoter is in plain italic; the *EarI* site used for linearization is in lower case; and the coding sequence for the RNA is bold).
2. 10× T7 transcription buffer: 400 mM Tris–HCl pH 8.0, 250 mM $MgCl_2$, 40 mM DTT, 20 mM spermidine.
3. 10× NTP mix: 20 mM ATP, 20 mM GTP, 20 mM CTP, and 5 mM UTP.
4. [α-^{32}P] UTP (6,000 Ci/mmol, 10 mCi/mL).
5. T7 RNA polymerase.
6. Temperature-controlled dry bath and heat block.
7. 2× RNA loading dye: 1× TBE (89 mM Tris, 89 mM boric acid, 2 mM EDTA, pH 8.3), 10 M urea, 10 mM EDTA, 0.002 % bromophenol blue, and 0.002 % xylene cyanol.
8. Electrophoresis system, including glass plates (20×20 cm), spacers and combs (typically 0.75 or 0.80 mm thick), gel-running apparatus, and a power supply.
9. Denaturing 15 % polyacrylamide gel solution (50 mL): 1× TBE, 7 M urea, 18.75 mL 40 % acrylamide (acrylamide:bis-acrylamide 29:1) stock solution.
10. Tetramethylethylenediamine (TEMED), 500 μL.
11. 10 % (w/v) ammonium persulfate (APS).
12. Autoradiography film.

13. 1× TEN buffer: 10 mM Tris–HCl pH 8.0, 1 mM EDTA, 100 mM NaCl.
14. Tube rotator.

2.4 Reconstituted pri-miRNA Processing Assays

1. RNaseOUT recombinant ribonuclease inhibitor.
2. Dry bath and electrophoresis system: Same as above.
3. Gelbond PAG film (Lonza).
4. Gel dryer, cold trap, and vacuum pump.
5. Storage phosphor screen.
6. Phosphorimager such as the Typhoon 9410 Variable Mode Imager (GE Healthcare).
7. Image analysis programs such as Quantity One (Bio-Rad), ImageQuant (GE Healthcare), or ImageJ (NIH, free online).

3 Methods

3.1 Expression of Heme-Bound DGCR8 NC1 in E. coli

1. Transform the NC1 expression plasmid to *E. coli*. Spread the bacteria on an LB agar plate with appropriate antibiotic and incubate at 37 °C overnight.
2. Inoculate a culture containing 150 mL of LB medium and appropriate antibiotic with a single colony. Shake at 250 rpm and 37 °C overnight.
3. Inoculate the desired volume of LB medium with antibiotic for overexpression. Shake at 250 rpm and 37 °C until $OD_{600\ nm}$ reaches 1.0–1.2 as measured using a spectrophotometer equipped with a turbidity cuvette holder (*see* **Note 3**).
4. Induce NC1 expression by adding IPTG and δ-ALA, both to a final concentration of 1 mM (*see* **Note 4**). Continue to shake for 3.5–4 h at the same temperature. Collect the cells by centrifuging at 5,000 × *g* at 4 °C for 15 min. The pellets should have a noticeable brown color.
5. Store the pellets at −80 °C.

3.2 Purification of NC1

1. Completely resuspend cell pellet in ice-cold DGCR8 lysis buffer (40 mL per L of culture).
2. Sonicate the cell suspension using the ultrasonic processor at 80 % power, 1-s on and 1-s off, for a total of 7–8 min. To avoid overheating the lysate, a 30-s break is taken after each minute of sonication and the container is kept on ice at all times.
3. Centrifuge the lysate at 45,000 × *g* for 30 min at 4 °C.
4. Load the supernatant onto a HiTrap SP column equilibrated with DGCR8 buffer *A*. Elute using a linear gradient of DGCR8 buffers *A* and *B*. The protein elutes at around 300 mM NaCl

(10 % DGCR8 buffer *B*). The fractions containing heme-bound NC1 have a yellowish-brown color. Analyze the purity of the fractions using sodium dodecyl sulfate-polyacrylamide gel electrophoresis (SDS-PAGE).

5. If the peak fractions are not >90 % pure, repeat the ion-exchange chromatography step. Pool the fractions containing relatively pure NC1. Dilute it 1:1 (v:v) with DGCR8 buffer *A* to reduce the salt concentration. Repeat **step 4**. The protein may be stored at 4 °C overnight if desired.
6. Equilibrate the Superdex 200 column in SEC buffer (roughly 50 mL). Concentrate the fractions from the last ion-exchange chromatography step down to ~550 μL using a centrifugal concentrator. Filter the concentrated NC1 solution through a membrane with 0.2 μm pores, and load the filtrate onto the column. Collect 0.5 mL fractions. Heme-bound NC1 dimer elutes at around 12.5 mL (*see* **Note 5**).
7. Determine the NC1 protein concentration using UV-visible absorption spectroscopy. Blank the spectrophotometer with SEC buffer. Scan between 240 and 700 nm. An absorption peak at 280 nm indicates that bacterial nucleic acids have been successfully removed from the protein. NC1 dimer concentration = $A_{280\ \mathrm{nm}}/\varepsilon_{280\ \mathrm{nm}}$. Based on the amino acid sequence, $\varepsilon_{280\ \mathrm{nm}}$ is estimated to be 94.5 $\mathrm{mM^{-1}\ cm^{-1}}$ ($\varepsilon_{280\ \mathrm{nm,apo}}$) [35]. This value has been used in all our previous publications. However, our recent unpublished measurements using microBCA assay indicate $\varepsilon_{280\ \mathrm{nm}} \approx 130\ \mathrm{mM^{-1}\ cm^{-1}}$ (*see* **Note 6**).
8. Calculate the $A_{450\ \mathrm{nm}}/A_{280\ \mathrm{nm}}$ ratio. If majority (>60 %) of the NC1 protein is occupied by heme, $A_{450\ \mathrm{nm}}/A_{280\ \mathrm{nm}}$ should be between 0.40 and 0.53 (the higher the better). NC1 preparations with lower $A_{450\ \mathrm{nm}}/A_{280\ \mathrm{nm}}$ ratios are less active (*see* **Notes 7** and **8**).
9. The Fe(III) heme-bound NC1 protein may be stored at 4 °C, with protection from light. Because this protein gradually loses pri-miRNA processing activity for reasons not well understood, we typically use it within a couple of days from the completion of purification.

3.3 Expression and Purification of Recombinant Homo Sapiens His$_6$-Drosha$^{390–1374}$

1. His$_6$-Drosha$^{390–1374}$ is expressed in Sf9 insect cells using a baculovirus system, following Invitrogen's standard protocols. The cell pellets are stored in −80 °C freezer until purification.
2. Resuspend a pellet from 50 mL of insect cell culture in 30 mL of ice-cold Drosha lysis buffer. Sonicate at 50 % power, 1-s on and 1-s off, for a total of 4 min. To avoid overheating the lysate, a 30-s break is taken after each minute of sonication and the container is kept on ice throughout the sonication procedure.
3. Centrifuge the lysate at 45,000 × *g* for 30 min at 4 °C.

4. Load the supernatant onto a Ni Sepharose High Performance column. Wash the column extensively using the Drosha wash buffer, and elute the His_6-Drosha$^{390-1374}$ protein in the Drosha elution buffer.
5. Dialyze the purified His_6-Drosha$^{390-1374}$ against the Drosha storage/reaction buffer.
6. Aliquot in 10 μL per tube, freeze in liquid nitrogen or dry ice–ethanol mix, and store in −80 °C freezer.

3.4 Transcription and Purification of pri-miRNAs Uniformly Labeled with ^{32}P

1. Set up the transcription reaction by adding the following:

   10 μL H_2O.

   2 μL 10× transcription buffer.

   2 μL 10× NTP mix.

   2 μL DNA template.

   2 μL [α-^{32}P] UTP.

   2 μL T7 RNA polymerase (always add last).

   Incubate the transcription reaction at 37 °C for 2–3 h.
2. Add 20 μL 2× RNA loading dye to the reaction.
3. Pour a denaturing 15 % polyacrylamide gel (1× TBE, 7 M urea). Assemble the gel sandwich. Induce polymerization of the 50 mL gel mix by adding 50 μL TEMED and 500 μL 10 % APS. Pour the gel, insert the comb, and let stand at room temperature for 1 h. Mount the gel on the electrophoresis apparatus. Pre-run the gel in 1× TBE at a constant power of 12 W for 20 min.
4. Load the transcription onto the gel. Run the gel at 12 W until the bromophenol blue reaches the bottom of the gel. Disassemble gel sandwich, and leave the gel attached to one glass plate.
5. Cover the gel with a plastic wrap. Expose the gel to an autoradiography film. Excise out the band. Expose the gel to another film to confirm the excision.
6. Crush and soak the gel piece in 1× TEN buffer at 4 °C for at least 10 h.
7. Precipitate the RNA by adding 3 volumes of ethanol and 0.1 volume of 3 M sodium acetate pH 5.2. Resuspend the RNA in H_2O.
8. Determine radioactivity using scintillation counting. Dilute the RNA to ~10,000 cpm/μL in water prior to the processing assay.

3.5 pri-miRNA Processing Assay

1. Set up a 10-μL processing reaction by mixing the following (in the order shown):

4.5 μL Drosha storage/reaction buffer.

2 μL recombinant His6-Drosha$^{390-1374}$ (~2 ng/μL).

1 μL Fe(III) heme-bound NC1 (10× stock).

0.5 μL RNaseOUT (40 U/μL).

1 μL 64 mM $MgCl_2$.

1 μL pri-miRNA (~10,000 cpm/μL).

For Drosha-only control, the NC1 stock solution should be replaced by SEC buffer. For NC1-only control, Drosha storage/reaction buffer should be used instead of the recombinant Drosha protein (*see* **Note 9**).

2. Incubate at 37 °C for 45 min. The reactions generally do not proceed further after 1 h. Add 10 μL 2× RNA loading dye to stop the reaction.
3. Analyze the reactions using a denaturing 15 % polyacrylamide gel (1× TBE, 7 M urea). Follow **steps 3** and **4** in Subheading 3.4, Transcription and purification of pri-miRNAs.
4. Adhere the disassembled gel on the hydrophobic surface of a Gelbond film (*see* **Note 10**). Cover the other side of the gel with a filter paper. Dry the gel using a gel dryer coupled to a cold trap and a vacuum pump.
5. Expose the gel to a storage phosphor screen overnight.
6. Scan in the image using a phosphorimager. An example image is shown in Fig. 3.
7. Quantify the total intensities of substrate and product bands using an image analysis program. Background intensities are subtracted. To calculate the fraction of pri-miRNA processed, the signals from the pre-miRNAs (*see* **Note 11**) are first converted to that of its corresponding pri-miRNA by multiplying the ratio of U residues in pri-miRNA and pre-miRNA, since the pri-miRNAs were uniformly labeled using [α-^{32}P] UTP. For example, the ratio is 42/16 = 2.625 for the 150-nt pri-miR-30a fragment as we used previously [22]. The signal of pri-miRNA processed is then divided by the amount of starting substrate.

4 Notes

1. The reducing reagent DTT (or β-mercaptoethanol) is important for keeping DGCR8 active. Make sure that your DTT stock solution is in a fully reduced state. Solutions of reduced DTT should have minimal absorbance above 250 nm, while oxidized DTT has an absorbance peak at 283 nm with an extinction coefficient of 273 M^{-1} cm^{-1} [36]. We store our DTT stock solution (1 M) in −20 °C in aliquots and avoid repeated freeze and thaw.

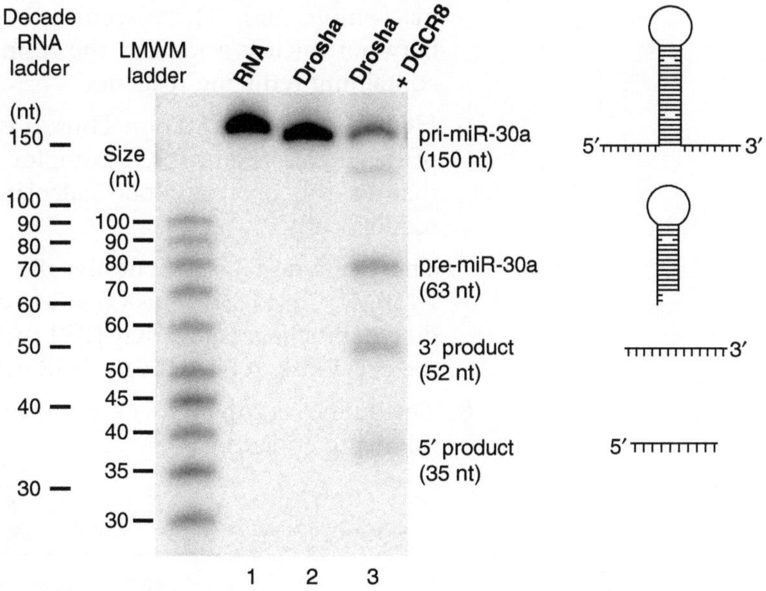

Fig. 3 Example of a pri-miRNA processing assay. Uniformly labeled pri-miR-30a was incubated with 4 nM His$_6$-Drosha$^{390-1374}$, either alone (*lane 2*) or with 50 nM Fe(III) heme-bound NC1 dimer (*lane 3*), at 37 °C for 45 min. The reactions were analyzed using a 7 M urea, 15 % polyacrylamide gel. LMWM: low molecular weight marker

2. The TEV cleavage site in His$_6$-Drosha$^{390-1374}$ allows the His$_6$-tag to be cleaved off using the TEV protease if desired but is not used in the protocol presented here.
3. Measuring OD$_{600\ nm}$ of cell cultures using a spectrophotometer with a turbidity cuvette holder is more accurate than without, because most of the scattered light is blocked by the turbidity cuvette holder. The absolute values of OD$_{600\ nm}$ depend on the instrument used. The bacterial cell density we use in NC1 expression is similar to that commonly recommended for protein expression.
4. δ-ALA is a key heme biosynthesis intermediate. In the absence of δ-ALA, NC1 is expressed as a mixture of heme-bound dimer and heme-free “monomer” [4]. Addition of δ-ALA increases the yield of NC1 expression and improves the heme content of NC1 to the extent that often little or no heme-free “monomer” is observed.
5. Presence of 1 mM DTT or 10 mM β-mercaptoethanol in the SEC buffer helps remove the residual amount of nucleic acids bound to NC1 during size exclusion chromatography using the Superdex 200 column. Under this condition, the free nucleic acids elute at >20 mL, a volume too large for

macromolecules. There seems to be an unusual interaction between nucleic acids and the resin in the presence of thiol-containing reducing reagents.

6. Heme absorbs at 280 nm. Thus, it is not surprising that $\varepsilon_{280\ \mathrm{nm}}$ of the Fe(III) heme–NC1 complex ($\varepsilon_{280\ \mathrm{nm,holo}}$) is higher than that of the protein alone calculated from the amino acid sequence ($\varepsilon_{280\ \mathrm{nm,apo}}$).
7. The extinction coefficient for the 450 nm peak of NC1 is 74 $\mathrm{mM^{-1}\ cm^{-1}}$ [30]. This value was determined using the pyridine hemochromagen assay [37] and should be used instead of the previously reported value (58 $\mathrm{mM^{-1}\ cm^{-1}}$) [4].
8. The heme occupancy (O_{heme}) may be calculated using the following equation:

$$O_{heme} = \frac{\varepsilon_{280,apo} \times \dfrac{A_{450}}{A_{280}}}{\varepsilon_{450} - \left(\varepsilon_{280,holo} - \varepsilon_{280,apo}\right) \times \dfrac{A_{450}}{A_{280}}},$$

where the $\varepsilon_{450\ \mathrm{nm}}$ of NC1 is 74 $\mathrm{mM^{-1}\ cm^{-1}}$ [30].

9. The above reaction can also be set up anaerobically using an anaerobic chamber to prepare the protein sample and a gastight syringe to transfer the pri-miRNA to the sample [6]. To do kinetic assay, increase the above volumes and take aliquots at desired intervals.
10. Polyacrylamide gels ≥15 % do not stick strongly to filter paper typically used in gel drying protocols but adhere to the hydrophobic surface of the Gelbond film. Note that this surface is opposite to the treated hydrophilic side that is designed to cross-link with acrylamide.
11. There is a low level of nonspecific nuclease activity associated with the His_6-Drosha$^{390-1374}$ protein, which may originate from either Drosha itself or residual impurities present in the Drosha preparation. The pre-miRNA is usually the most stable among the three specific Drosha cleavage products and thus is used for quantification.

Acknowledgments

We thank Jen Cleveland for comments. This work was supported by the National Institute of General Medical Sciences of the National Institutes of Health under award number R01GM080563 (F.G.) and by a UCLA Dissertation Year Fellowship (I.B.).

References

1. Lee Y, Ahn C, Han J et al (2003) The nuclear RNase III Drosha initiates microRNA processing. Nature 425:415–419
2. Denli AM, Tops BB, Plasterk RH et al (2004) Processing of primary microRNAs by the microprocessor complex. Nature 432: 231–235
3. Gregory RI, Yan KP, Amuthan G et al (2004) The microprocessor complex mediates the genesis of microRNAs. Nature 432:235–240
4. Faller M, Matsunaga M, Yin S et al (2007) Heme is involved in microRNA processing. Nat Struct Mol Biol 14:23–29
5. Barr I, Smith AT, Senturia R et al (2011) DiGeorge critical region 8 (DGCR8) is a double-cysteine-ligated heme protein. J Biol Chem 286:16716–16725
6. Barr I, Smith AT, Chen Y et al (2012) Ferric, not ferrous, heme activates RNA-binding protein DGCR8 for primary microRNA processing. Proc Natl Acad Sci U S A 109:1919–1924
7. Wang Y, Medvid R, Melton C et al (2007) DGCR8 is essential for microRNA biogenesis and silencing of embryonic stem cell self-renewal. Nat Genet 39:380–385
8. Yi R, Pasolli HA, Landthaler M et al (2009) DGCR8-dependent microRNA biogenesis is essential for skin development. Proc Natl Acad Sci U S A 106:498–502
9. Faller M, Guo F (2008) MicroRNA biogenesis: there's more than one way to skin a cat. Biochim Biophys Acta 1779:663–667
10. Han J, Lee Y, Yeom KH et al (2004) The Drosha-DGCR8 complex in primary microRNA processing. Genes Dev 18:3016–3027
11. Fukuda T, Yamagata K, Fujiyama S et al (2007) DEAD-box RNA helicase subunits of the Drosha complex are required for processing of rRNA and a subset of microRNAs. Nat Cell Biol 9:604–611
12. Guil S, Caceres JF (2007) The multifunctional RNA-binding protein hnRNP A1 is required for processing of miR-18a. Nat Struct Mol Biol 14:591–596
13. Viswanathan SR, Daley GQ, Gregory RI (2008) Selective blockade of microRNA processing by Lin28. Science 320:97–100
14. Davis BN, Hilyard AC, Lagna G et al (2008) SMAD proteins control DROSHA-mediated microRNA maturation. Nature 454:56–61
15. Trabucchi M, Briata P, Garcia-Mayoral M et al (2009) The RNA-binding protein KSRP promotes the biogenesis of a subset of microRNAs. Nature 459:1010–1014
16. Suzuki HI, Yamagata K, Sugimoto K et al (2009) Modulation of microRNA processing by p53. Nature 460:529–533
17. Sakamoto S, Aoki K, Higuchi T et al (2009) The NF90-NF45 complex functions as a negative regulator in the microRNA processing pathway. Mol Cell Biol 29:3754–3769
18. Michlewski G, Caceres JF (2010) Antagonistic role of hnRNP A1 and KSRP in the regulation of let-7a biogenesis. Nat Struct Mol Biol 17:1011–1018
19. Zeng Y, Cullen BR (2005) Efficient processing of primary microRNA hairpins by Drosha requires flanking nonstructured RNA sequences. J Biol Chem 280:27595–27603
20. Yeom KH, Lee Y, Han J et al (2006) Characterization of DGCR8/Pasha, the essential cofactor for Drosha in primary miRNA processing. Nucleic Acids Res 34:4622–4629
21. Sohn SY, Bae WJ, Kim JJ et al (2007) Crystal structure of human DGCR8 core. Nat Struct Mol Biol 14:847–853
22. Faller M, Toso D, Matsunaga M et al (2010) DGCR8 recognizes primary transcripts of microRNAs through highly cooperative binding and formation of higher-order structures. RNA 16:1570–1583
23. Guo F (2012) Drosha and DGCR8 in microRNA biogenesis. In: Guo F, Tamanoi F (eds) The enzymes: eukaryotic RNases and their partners in RNA degradation and biogenesis, Part B, vol 32, 1st edn. Elsevier Academic Press, Amsterdam, Netherlands, pp. 101–121
24. Lee Y, Jeon K, Lee JT et al (2002) MicroRNA maturation: stepwise processing and subcellular localization. EMBO J 21:4663–4670
25. Newman MA, Thomson JM, Hammond SM (2008) Lin-28 interaction with the Let-7 precursor loop mediates regulated microRNA processing. RNA 14:1539–1549
26. Lee Y, Kim VN (2007) In vitro and in vivo assays for the activity of Drosha complex. Methods Enzymol 427:89–106
27. Zeng Y, Yi R, Cullen BR (2005) Recognition and cleavage of primary microRNA precursors by the nuclear processing enzyme Drosha. EMBO J 24:138–148
28. Zhang X, Zeng Y (2010) The terminal loop region controls microRNA processing by Drosha and Dicer. Nucleic Acids Res 38: 7689–7697
29. Senturia R, Faller M, Yin S et al (2010) Structure of the dimerization domain of DiGeorge Critical Region 8. Protein Sci 19: 1354–1365
30. Senturia R, Laganowsky A, Barr I et al (2012) Dimerization and heme binding are conserved in amphibian and starfish homologues of the microRNA processing protein DGCR8. PLoS One 7(7):e39688. doi:10.1371/journal.pone.0039688

31. Kim VN, Han J, Siomi MC (2009) Biogenesis of small RNAs in animals. Nat Rev Mol Cell Biol 10:126–139
32. Morlando M, Ballarino M, Gromak N et al (2008) Primary microRNA transcripts are processed co-transcriptionally. Nat Struct Mol Biol 15:902–909
33. Kim YK, Kim VN (2007) Processing of intronic microRNAs. EMBO J 26:775–783
34. Milligan JF, Uhlenbeck OC (1989) Synthesis of small RNAs using T7 RNA polymerase. Methods Enzymol 180:51–62
35. Pace CN, Vajdos F, Fee L et al (1995) How to measure and predict the molar absorption coefficient of a protein. Protein Sci 4: 2411–2423
36. Cleland WW (1964) Dithiothreitol, a new protective reagent for SH groups. Biochemistry 3:480–482
37. Berry EA, Trumpower BL (1987) Simultaneous determination of hemes a, b, and c from pyridine hemochrome spectra. Anal Biochem 161:1–15

Chapter 6

In Vivo Processing Assay Based on a Dual-Luciferase Reporter System to Evaluate DROSHA Enzymatic Activity

Vera Bilan, Danilo Allegra, Florian Kuchenbauer, and Daniel Mertens

Abstract

Luciferase reporter assays are widely used to study promoter activity, transcription factors, intracellular signaling, protein interactions (Jia et al., PloS One 6:e26414), miRNA processing (Allegra and Mertens, Biochem Biophys Res Commun 406:501–505), and target recognition (Jin et al., Methods Mol Biol 936:117–127). Here we describe the use of a dual-luciferase reporter system to evaluate the enzymatic activity of a key enzyme involved in RNA maturation—DROSHA. This dual system is a simple and fast method for the quantification of the DROSHA processing activity in live cells.

Key words MicroRNA, Luciferase, miRNA processing assay, DROSHA, Ribonuclease III

1 Introduction

Luciferase reporter assays are widely used to study promoter activity, transcription factors, intracellular signaling, protein interactions [1], miRNA processing [2], and target recognition [3]. A dual-reporter assay using firefly (*Photinus pyralis*) and *Renilla* (*Renilla reniformis*, also known as sea pansy) luciferases improves experimental accuracy by normalizing the results and reducing technical variance. Firefly (FF) and Renilla luciferases do not require posttranslational modifications and function as an enzymatic reporter immediately after translation. Additionally, these two luciferases have dissimilar substrate requirements making it possible to measure them subsequently from a single sample [4]. In dual-reporter assays Renilla luciferase is normally used as an internal control to reduce experimental variability [5]. We adapted a protocol that applies a dual-luciferase assay to measure enzymatic activity of the DROSHA enzyme.

DROSHA is a member of the ribonuclease III family of proteins and catalyzes cleavage of the primary long miRNA transcript (pri-miRNA) to its precursor miRNA form. The process is critical for miRNA maturation although a small subgroup of miRNA molecules are processed without this step. DROSHA has two RNase

Christoph Arenz (ed.), *miRNA Maturation: Methods and Protocols*, Methods in Molecular Biology, vol. 1095, DOI 10.1007/978-1-62703-703-7_6, © Springer Science+Business Media New York 2014

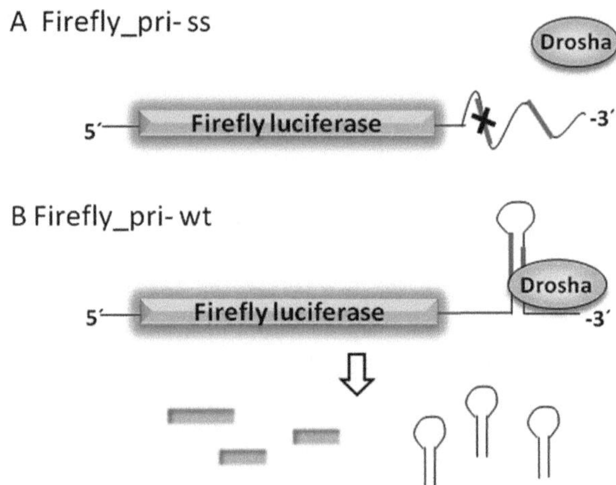

Fig. 1 Schematic representation of luciferase-based measurement of DROSHA processing activity

III domains and one double-stranded RNA-binding domain. The most important cofactor is a DiGeorge critical region 8 protein (DGCR8, also known as Pasha) that forms a microprocessor complex with DROSHA [6].

It is believed that DROSHA does not possess sequence specificity. Instead it recognizes the RNA secondary structure. Pri-miRNAs that are recognized by DROSHA have a hairpin structure with a large, unstructured terminal loop and a stem of more than 26 base pairs (bp) but are less than 40 bp in length [7]. DROSHA preferentially cleaves 11 bp away from free ends of pri-mRNA at the base of the hairpin stem [8].

In our assay we used this recognition feature of DROSHA enzyme to quantify its enzymatic activity in live cells. As an exemplary pri-miR, we chose pri-16-1 because its recognition and processing by the microprocessor complex have been extensively studied. We constructed two plasmids: one plasmid contains the hairpin fragment of the pri-miRNA to be analyzed with an additional 25 bp on each side of the hairpin. This wild-type (wt) sequence is cloned at the 3′UTR of the firefly gene. Here we used the pri-miRNA-16-1 sequence. The second plasmid is used as a control plasmid and contains the same sequence (here of pri-miRNA-16-1) but with four point mutations that were introduced into the hairpin stem. These mutations destabilize the hairpin structure. This control plasmid can therefore not be recognized and cleaved by DROSHA and serves as a positive normalization control for sequence-dependent and sequence-independent effects of the 3′UTR (*see* Fig. 1). Both constructs should have a similar stability [2].

2 Materials

2.1 Cell Culture

Medium should be stored at 4 °C and warmed up to 37 °C before use.

1. Dulbecco's modified Eagle's medium (DMEM).
2. Fetal bovine serum (FBS).
3. 1× Phosphate-buffered saline (PBS, pH 7.4).
4. 5× Trypsine. Dilute to 1× solution in PBS and store at 4 °C.
5. HEK293T cells maintained in DMEM supplemented with 10 % FBS at 37 °C with 10 % CO_2 in 75 cm^2 flasks.

2.2 Transfection Reagents

1. TransIT®-LT1 Transfection Reagent (Mirus Bio LLC, USA). Store at −20 °C.

2.3 Plasmids

1. pMir-Report Luciferase (Ambion, Austin, USA).
2. pSuper (Oligoengine).
3. pRL-CMV (Promega Inc., USA).
4. pMir-Report Luciferase pri-miR-16-1wt (amplified from genomic DNA with primers (for sequences *see* Subheading 2.5, Table 2) and cloned with Hind III into pMir-Report Luciferase).
5. pMir-Report Luciferase pri-miR-16-1ss ("ss" meaning "single stranded"; mutated using primers (for sequences *see* Subheading 2.5, Table 2) and cloned with Hind III into pMir-Report Luciferase).
6. pSuper_shEGFP (synthetic oligos (*see* Subheading 2.4, Table 1) were cloned between the BglII and HindIII sites of pSuper plasmid, Oligoengine).
7. pSuper_shDROSHA (synthetic oligos (*see* Subheading 2.4, Table 1) were cloned between the BglII and HindIII sites of pSuper plasmid, Oligoengine).

2.4 Oligos for shRNA Plasmids

Table 1
shRNA sequences used for cloning into pSuper plasmids

Name	Sequence (oligoengine)
shDROSHA_s	5′-GATCCCCGCGAGTAGGCTTCGTGACTTATATGACACCTGACCCATGTCATATAAGTCACGAAGCCTACTCGTTTTTA-3′
shDROSHA_as	5′-AGCTTAAAAACGAGTAGGCTTCGTGACTTATATGACATGGGTCAGGTGTCATATAAGTCACGAAGCCTACTCGCGGG-3′
shEGFP_s	5′-GATCCCCGCTGACCCTGAAGTTCATCTTCAAGAGAGATGAACTTCAGGGTCAGCTTTTTA-3′
shEGFP_as	5′-AGCTTAAAAAGCTGACCCTGAAGTTCATCTCTCTTGAAGATGAACTTCAGGGTCAGCGGG-3′

2.5 Primers for Cloning

Table 2
Primers for cloning pri-miRNA into pMir-Report Luciferase plasmids

Primer name	Sequence
pri-miR-16-1 FOR	5′-TGATAGCAATGTCAGCAGTG-3′
pri-miR-16-1 REV	5′-TGGTCAACCTTACTTCAGCA-3′
Mutagenesis	
pri-miR-16-1 ss_f	5′-TACCACCAGGTAAAAATTGGCGTTAAG ATTCTAAAATTATC TCC-3′
pri-miR-16-1 ss_r	5′-CGCCAATTTTTACCTGGTGGTAAGGCA CTGCTGACATTGCT-3′

2.6 In Vivo Processing Assay

Luciferase buffers should be aliquoted and stored at −20 °C. Repeated freezing–thawing cycles of these buffers may decrease assay performance.

1. 5× Passive Lysis Buffer (Promega). Store at −20 °C. 1× solution in ddH_2O can be stored at 4 °C for up to 1 month.
2. Firefly buffer pH (8.0): 25 mM glycylglycine, 15 mM K_xPO_4 (pH 8.0), 4 mM EGTA, 2 mM ATP, 1 mM DTT, 15 mM $MgSO_4$, 0.1 mM CoA, 75 μL luciferin (*see* **Notes 1** and **2**).
3. Renilla buffer pH (5.0): 1.1 M NaCl, 2.2 mM Na_2EDTA, 0.22 M K_xPO_4 (pH 5.1), 0.44 mg/mL bovine serum albumin (BSA), 1.3 mM NaN_3, 1.43 μM coelenterazine.
4. Luminometer, like GloMax 96 (Promega).

3 Methods

3.1 Transfection of Adherent Cells

3.1.1 Knockdown of DROSHA Using shRNA

1. One day before transfection seed the cells into two 75 cm^2 flasks. Cell density should reach 50–80 % confluence on the day of transfection.
2. The day of transfection, replace the medium with 9 mL of complete growth medium.
3. Allow the transfection reagent to warm up to room temperature. Pre-warm serum-free DMEM medium to 37 °C.
4. For each flask to be transfected prepare a *transfection mix* by mixing 8 μg pSUPER-shEGFP (*see* Table 1 for sequences or 8 μg pSUPER-shDROSHA (*see* Table 1)) with 980 μL serum-free DMEM. Add 20 μL TransIT-LT1 transfection reagent, mix well, and incubate for 20 min at room temperature (*see* **Note 3**).
5. Dropwise add 1 mL of transfection complexes to the cells and shake gently by rocking the plate back and forth.

6. Four days after transfection proceed with in vivo assay (*see* **Note 4**). Complexes do not have to be removed following transfection (*see* **Note 5**).

3.1.2 In Vivo Assay Transfection

1. One day before transfection seed the cells into 12-well plates. Cell density should reach 50–80 % confluence on the day of transfection.
2. On the day of transfection, replace the medium with 1 mL of complete growth medium.
3. For each well to be transfected prepare *transfection mix* by mixing 200 pRL-CMV+100_ng pMirLuc16-2ss+1 μg_ pSuper-shEGFP in 93.5 μL serum-free DMEM (*see* **Note 6**). Add 4 μL TransIT-LT1 transfection reagent, mix well, and incubate for 20 min at room temperature.
4. Dropwise add 100 μL of transfection mix to the cells and shake gently by rocking the plate back and forth.
5. Lyse cells and assay luciferase signal 18–24 h after transfection.

3.2 Cell Lysis

1. Prepare proper amount of 1× passive lysis buffer by diluting one part of 5× passive lysis buffer (keep on ice) in four parts of ddH_2O (*see* **Note** 7).
2. Gently remove the medium from the cell culture wells and rinse cells with 500 μL 1× PBS.
3. Add 100 μL of 1× passive lysis buffer and shake at room temperature for 10 min.
4. Transfer the lysate to a 1.5-mL microcentrifuge tube, and spin down cells at 16,000×*g* at room temperature for 1 min to remove cell debris.
5. Transfer supernatant into a new 1.5-mL microcentrifuge tube, and keep the samples on ice before measurement (*see* **Note 8**).
6. Pipet 15 μL of each sample into 96-well plate (*see* **Note 9**).

3.3 Measurement of Luciferase Activity

1. Thaw proper amount of buffers referring to 100 μL buffer/ well. Let warm to room temperature. Always mix the reagent prior to use as thawing generates density and composition gradients (*see* **Notes 10** and **11**).
2. For measurement a luminometer can be used with two automatic injectors. The following parameters are set in our instrument settings: injection volume of buffer 75 μL with 5-s delay between injection and measurement (*see* **Note 12**).
3. Depending on the luminometer used it might be necessary to prime auto-injector systems with buffers before measurement.
4. Record the firefly luciferase activity measurement by adding FF buffer to the well.

5. Record the Renilla luciferase activity measurement by adding Renilla buffer to the well.

3.4 Evaluation of Results

1. Luciferase assay results are normalized as follows:

$$Normalized\ activity = \frac{Firefly\ luciferase\ activity}{Renilla\ luciferase\ activity}$$

2. The processing efficiency of pri-miRNA construct can be calculated:

$$pri\text{-}miRNA\ processing\ efficiency = 1 - \frac{normalized\ activity\ of\ wt\ construct}{normalized\ activity\ of\ ss\ construct}$$

In this calculation the processing efficiency of the ss construct will be equal to 0.

4 Notes

1. Dithiothreitol (DTT) is highly unstable in water solutions, aliquot and store at −20 °C, and thaw the aliquots on ice.
2. The addition of Coenzyme A (CoA) allows a more sustained plateau of activity facilitating subsequent analysis. CoA in micromolar concentrations produces a more intense and sustained light emission because of the more favorable kinetics of the reaction of the enzyme with the luciferyl CoA substrate. In the presence of CoA, the luciferase assay yields stabilized luminescence signals with significantly greater intensities.
3. Transfection reagent should be added at the end of the master mix preparation.
4. The reaction times can vary for different cell lines. The protocol should be optimized before starting the experiments.
5. Depending on the reagent used for transfection it might be necessary to change the medium at different time points. Follow the manufacturer's instructions when transecting cells with a different reagent.
6. The amount of plasmids and reagent used for the experiments needs to be optimized for every cell type.
7. Presence of detergents in the lysis buffer (i.e., TritonX) contributes to increase in the intensity and duration of light emission. However, some types of nonionic detergents commonly used to prepare cell lysates (i.e., Triton X-100) can intensify coelenterazine autoluminescence.
8. The stability of FF luc has been shown to be 6 weeks at 4 °C in cell lysates.

9. The amount of extract required may vary depending on the luciferase expression level and the instrumentation used; the amount used should be adjusted to keep the signal within the linear range of the assay.
10. At low pH the wavelength of emitted light from the FF reaction is shifted to lower wavelengths.
11. Control of temperature is also important since the use of non-equilibrated reagents directly from the refrigerator can cause a 5–10 % decrease in efficiency.
12. Depending on the luminometer used it might be necessary to prime auto-injector systems with buffers. Calculate additional volume of the buffers for this.

References

1. Jia S, Peng J, Gao B et al (2011) Relative quantification of protein-protein interactions using a dual luciferase reporter pull-down assay system. PLoS One 6:e26414
2. Allegra D, Mertens D (2011) In-vivo quantification of primary microRNA processing by Drosha with a luciferase based system. Biochem Biophys Res Commun 406:501–505
3. Jin Y, Chen Z, Liu X et al (2013) Evaluating the MicroRNA targeting sites by luciferase reporter gene assay. Methods Mol Biol 936: 117–127
4. Bronstein I (1994) Chemiluminescent and bioluminescent reporter gene assays. Anal Biochem 169–181
5. Stables J, Scott S, Brown S et al (1999) Development of a dual glow-signal, firefly and renilla luciferase assay reagent for the analysis of G-protein coupled receptor signalling. J Recept Signal Transduct Res 19:395–410
6. Winter J, Jung S, Keller S et al (2009) Many roads to maturity: microRNA biogenesis pathways and their regulation. Nat Cell Biol 11:228–234
7. Zeng Y, Yi R, Cullen BR (2005) Recognition and cleavage of primary microRNA precursors by the nuclear processing enzyme Drosha. EMBO J 24:138–148
8. Han J, Lee Y, Yeom K-H et al (2006) Molecular basis for the recognition of primary microRNAs by the Drosha-DGCR8 complex. Cell 125:887–901

Chapter 7

Assaying Dicer-Mediated miRNA Maturation by Means of Fluorescent Substrates

Marlen Hesse, Brian P. Davies, and Christoph Arenz

Abstract

Assaying Dicer-mediated miRNA maturation is a valuable tool not only for validating miRNA maturation itself but also for testing Dicer activity in cell lysate and for screening small molecules inhibiting miRNA maturation in a high-throughput format. The classical assay for miRNA maturation relies on radioactive labeling of a pre-miRNA and subsequent gel electrophoresis and autoradiography. Here we present a fluorescently labeled and quenched pre-miRNA beacon that can be ligated easily out of two single labeled RNA strands. Upon Dicer cleavage of the beacon, fluorophore and quencher are separated, which results in an increase of fluorescence over time. Unlike ^{32}P-labeled probes, our fluorescently labeled pre-miRNA beacon is stable for at least 5 years under storage conditions. Dicer or miRNA maturation assays can be easily performed in a 384-well plate format, consuming less than 1 pmol of RNA beacon per reaction.

Key words miRNA, Fluorescence assay, Nonradioactive assay, Dicer cleavage, miRNA maturation, miRNA inhibition

1 Introduction

As abnormal expression of miRNA has been found to influence—and obviously also induce—human disease, the analysis and manipulation of miRNA maturation processes need easy-to-use assays for further investigation [1, 2]. Radioactive Dicer assays [3] or miRNA maturation assays [4, 5] are tedious because of electrophoretic separation and autoradiography after labeling and enzyme reaction. Moreover, short ^{32}P half-life requires repetitive labeling, whereas appropriately stored RNA beacon is stable in our hands for at least 5 years (*see* **Note 1**).

In the special case of evaluating small-molecule inhibitors of miRNA maturation, multiplex reactions including variation of concentration are necessary and a homogenous assay method is required.

Christoph Arenz (ed.), *miRNA Maturation: Methods and Protocols*, Methods in Molecular Biology, vol. 1095, DOI 10.1007/978-1-62703-703-7_7, © Springer Science+Business Media New York 2014

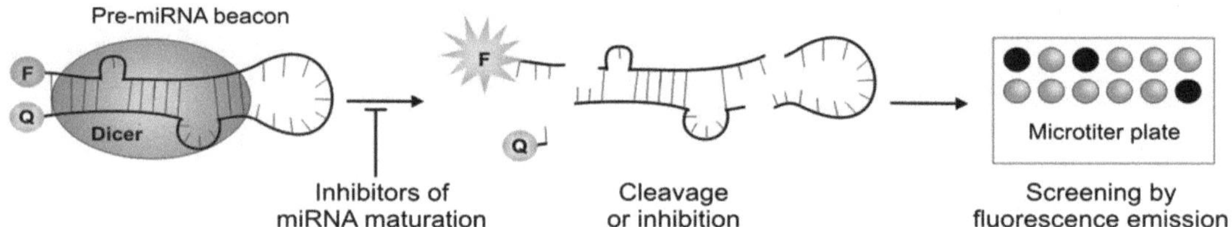

Fig. 1 Homogenous assay of miRNA maturation. Native structure of pre-miRNA beacon is cleaved by Dicer enzyme. Cleavage cuts off quencher and fluorophore leading to fluorescent increase. In case of potent inhibitors, Dicer cleavage is blocked and quenching is still on

Monitoring the Dicer-mediated miRNA maturation step by this homogenous assay based on a fluorescent substrate has great potential: (1) Any desired pre-miRNA can be labeled with fluorophore and quencher to validate miRNA maturation using recombinant Dicer enzyme or cell lysate. (2) Assay can be used to determine Dicer activity of cell lysate with respect to one specific miRNA. (3) Potential inhibitors of miRNA maturation can be tested in the manner of high-throughput screen using recombinant Dicer enzyme [6, 7]. The homogenous assay of miRNA maturation (Fig. 1) is based on a fluorescent pre-miRNA probe that contains a 5′-fluorophore (F) and a 3′-quencher (Q). Before cleavage, fluorescence is quenched because of the close proximity of fluorophore and quencher in the native hairpin formation of pre-miRNA. Dicer cleaves that native structure and enables the fluorophore to emit its fluorescence without quenching, which can be detected at the proper wavelength. In the presence of an inhibitor that binds to pre-miRNA, Dicer cleavage is blocked and quenching persists. Thus, a lower or no increase in fluorescence is observed. Especially for screening purposes this assay can be used in 384-well microplates using a plate reader (*see* **Note 2**).

It should be mentioned that Dicer usually recognizes its substrates via interaction with the specific 5′-phosphate and 3′- two nucleotide overhang at the terminus of the pre-miRNA stem [8]. Cleavage specificity then is most probably set by the spatial distance of the dsRNA from the specific recognition site. The lack of this specific signature may lead to a shifted cleavage of pre-miRNA [9], which cannot be excluded for end-labeled pre-miRNA as described here. Such a shift in the cleavage site is irrelevant for the evaluation of relative Dicer activity, but may affect the validity of results, when pre-miRNA binders as potential inhibitors of miRNA maturation are screened. However, by directly comparing results for such inhibitor testings using this assay and an alternative assay relying on unmodified pre-miRNA, we found no significant differences [10]. Recently, similar assays for Dicer activity [11] or miRNA maturation [12] have been described.

2 Materials

Use all reagents of the highest quality available and certified pyrogen/DNAse/RNAse free when possible. Prepare and dilute all solutions using ultrapure water (prepared by purifying deionized water to attain a sensitivity of 18 MΩ cm at 25 °C). Diligently follow all waste disposal regulations when disposing waste materials.

2.1 Ligation of 5′- and 3′-Terminal Partial Strands and Beacon Purification

1. Labeled RNA strands: 5′-FAM 36 mer and 3′-DABCYL 36 mer (*see* **Note 1**).
2. T4 RNA ligase 1 supplied with 10× T4 RNA ligase reaction buffer and 10 mM ATP (New England Biolabs).
3. RNase Inhibitor.
4. PCI extraction mix: Phenol/chloroform/isoamyl alcohol 25:24:1, pH 7.5–8.0.
5. 0.1 M Triethylammonium acetate (TEAA) buffer: TEAA, pH 7.0.

2.2 Analysis of Ligation Product

1. SYBR Gold (Invitrogen).
2. 0.3 M 2,4,6-trihydroxyacetophenone (THAP) solution, in ethanol.
3. 0.1 M Diammonium citrate solution (pH 6.4), in water.

2.3 Preparation of Pre-miRNA Prior to Fluorescence Assaying

1. Recombinant Dicer enzyme (Genlantis, San Diego, CA, USA).
2. Dicer buffer (1×): 20 mM Tris–HCl, pH 6.8, 12.5 mM NaCl, 2.5 mM $MgCl_2$, 1 mM DTT (*see* **Note 3**).

2.4 Fluorescence Assay

See Subheading 2.3.

3 Methods

3.1 Ligation of 5′- and 3′-Terminal Partial Strands and Beacon Purification

1. Ligate two partial pre-miRNA strands with FAM- and DABCYL-modified ends (Fig. 2).
2. For a 10 nmol reaction prepare ligation mixture with final concentration of 40 μM 5′-FAM 36 mer, 40 μM 3′-DABCYL 36 mer, 1× T4 Ligase buffer, 1 mM ATP, 0.8 U/μl RNase inhibitor, 0.8 U/μl T4 RNA ligase 1. Reaction time is 18 h at 37 °C.
3. Purify RNA by PCI extraction and precipitation according to standard procedure [13] followed by RP-HPLC (*see* **Note 4**). Use a gradient from 3 to 40 % 0.1 M TEAA in acetonitrile with flow rate of 1 ml/min over 40 min.

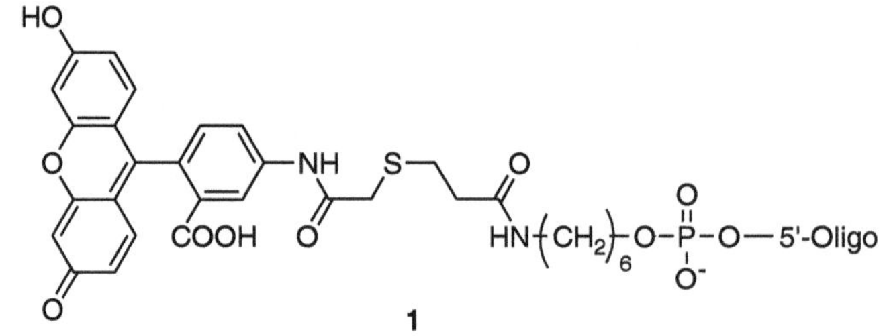

Fig. 2 Structure of 5′- and 3′-terminal ends. RNA beacon is formed by ligating two 36 mers, one with the 5′-fluorescein modification (FAM-EX-5 linker, **1**) and the other with the 3′-Dabcyl linker containing a 5′-phosphate (**2**). Linkers shown are purchased from IBA (Germany) and may vary by supplier

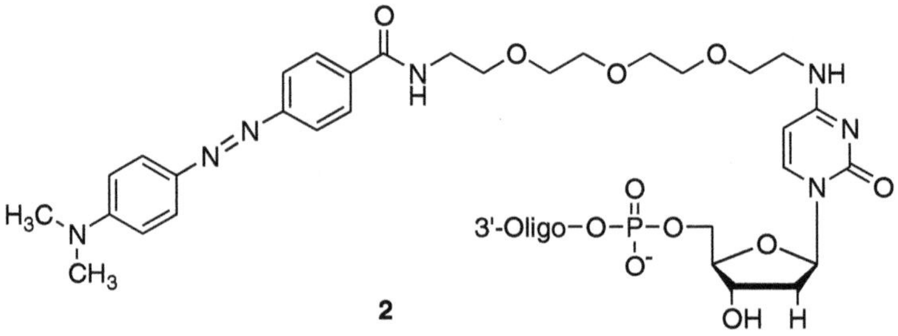

4. Analyze by MALDI-TOF and PAGE. Successful ligation leads to the 5′-FAM- and 3′-DABCYL-labeled pre-miRNA beacon.
5. Extinction coefficients for RNA were calculated using the nearest neighbor method adding extinction coefficients for fluorophore and quencher. Pure fractions are lyophilized and stored in 1 mM ammonium citrate, pH 6.4.

3.2 Analysis of Ligation Product

1. Check for adequate ligation by polyacrylamide gel electrophoresis (PAGE). Use 20 % denaturing 8 M urea PAGE gels as well as native PAGE gels in TBE buffer according to standard procedure [13].
2. Stain the gels with SYBR Gold. On a native gel without staining the successful quenching of the ligation product can be observed with ex = 491 nm (*see* **Note 5**). After staining with SYBR Gold the quenched pre-miRNA beacon can be visualized (Fig. 3).
3. Perform MALDI-TOF measurements to evaluate the ligation product. Settings for RNA consist of a linear and negative ion mode.
4. Prepare a matrix THAP/citrate in a ratio of 2:1 v/v from 0.3 M THAP in ethanol and 0.1 M diammonium citrate (pH 6.4) in water.

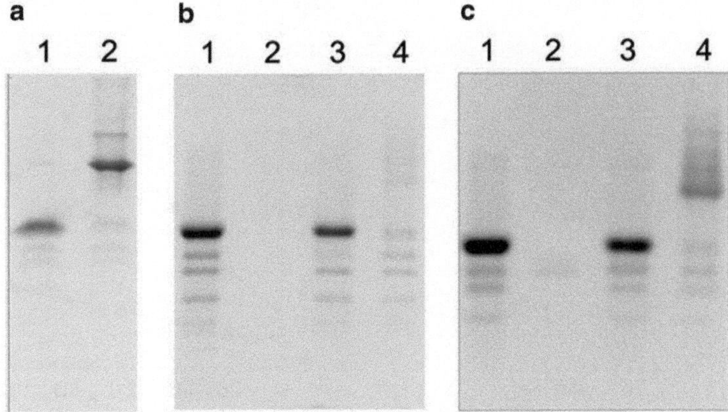

Fig. 3 Analysis of ligation product (pre-miRNA beacon) with 20 % gels. (**a**) Denaturing PAGE of ligation reaction with the mixed fragments (*1*) and the ligation product (*2*). (**b**) Native PAGE of ligation without staining. FAM-36-mer (*1*), DABCYL-36-mer (*2*), mixed fragments (*3*), ligation product (*4*). (**c**) Same as in (**b**) but stained with fluorescent dye which detects ligation product (*4*) that is quenched

3.3 Preparation of Pre-miRNA Prior to Fluorescence Assaying

1. Prior to use in the assay de- and renature labeled pre-miRNA at a stock concentration of 200 nM in 1× Dicer buffer by heating for 2 min at 95 °C and cooling to room temperature (rt) for over 30 min.
2. Test if Dicer can actually cleave prepared pre-miRNA beacon (*see* **Note 6**).

3.4 Fluorescence Assay

1. For Dicer cleavage assay prepare reaction mixture in one well in a 384-well microtiter plate on ice. Reaction mixture with total volume of 40 μl per well contains 20 nM labeled pre-miRNA, 1× Dicer reaction buffer, and 2.5 U/ml Dicer enzyme or heat-denatured Dicer for negative control (*see* **Note 7**).
2. Place the plate into plate reader, and measure fluorescence increase every 1–10 min for 4 h with ex = 491 nm, em = 520 nm, and gain = 2,500 (*see* **Note 8**).
3. Analyze fluorescence increase from the slope of the linear region that occurs in the range of 5–50 min (Fig. 4). For avoiding false-positive results, do a control using RNase inhibitor (*see* **Note 9**).
4. For testing of potential pre-miRNA binders (test substances) combine 32 μl of a 25 nM pre-miRNA solution in 1× Dicer buffer with 4 μl test substance in one well in a 384-well microtiter plate.
5. Cover with parafilm, and incubate for 30 min at rt.

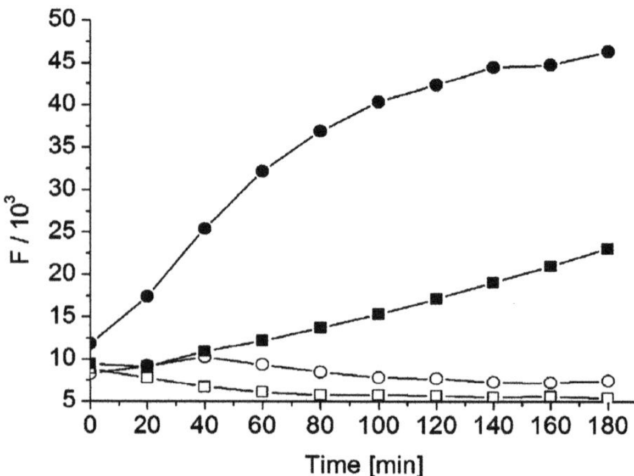

Fig. 4 Example graph from the assay in a 384-well format. Fluorescence increase upon incubation of 0.5 U/ml Dicer with 20 nM pre-miRNA beacon (*filled circle*). In the presence of an inhibitor (graph shows 100 μM kanamycin A, *filled square*) increase is diminished. Controls are heat-denatured Dicer with (*open square*) or without (*open circle*) inhibitor, which do not show any increase over the indicated amount of time

6. After incubation, place plate on ice for 10 min before adding 4 μl of Dicer enzyme in 1× Dicer buffer with final concentration of 2.5 U/ml. Final pre-miRNA concentration is 20 nM.
7. Keep on ice until placing plate into plate reader and measure fluorescence increase as described above.

4 Notes

1. Labeled RNA strands were purchased from IBA (Göttingen, Germany). Local suppliers may use other chemical linkers than those shown in Fig. 2 for the fluorescein/DABCYL modifications, which probably will not compromise the feasibility of the assay.
2. This fluorescence assay or rather the labeled pre-miRNA beacon may be considered candidate for real-time miRNA maturation assay and Dicer cleavage assay, respectively, in living cells.
3. Salt concentration considerably influences Dicer activity. For measuring inhibition of miRNA maturation through binding to pre-miRNA with certain test substances, physiologic salt concentration is recommended. In contrast to low-salt conditions, physiologic salt concentration may not be the optimum for Dicer activity, but outcome reveals more realistic and closer-to-cell values.

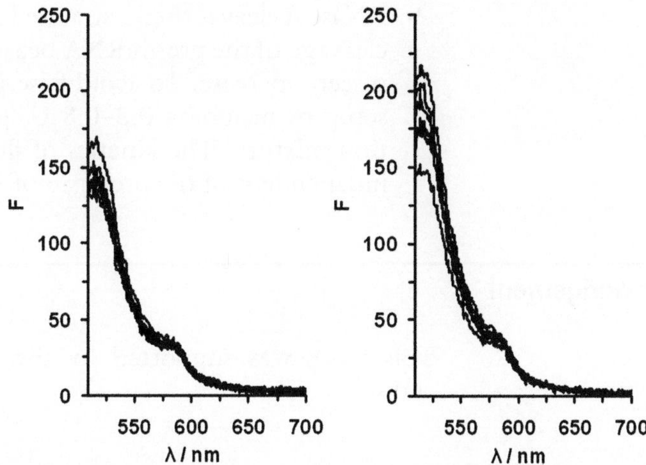

Fig. 5 Cleavage with Dicer. Dicer-mediated fluorescence increase was measured over the course of 4 h. Incubation of 100 nM beacon with 20 U/ml heat-denatured Dicer (*left*) and with 20 U/ml recombinant Dicer (*right*)

4. For RP-HPLC (reverse-phase HPLC) we used a "Polaris" column from Varian Inc. (5 μm, 200 Å, 250 mm × 4.6 mm) heated at 50 °C. You should pay attention on working RNase free as far as possible during and after HPLC purification.
5. Before analysis on a native gel, the pre-miRNA beacon should be de- and renatured to form its proper secondary structure, which enables perfect quenching due to the close proximity of fluorophore and quencher. The lower this background signal appears, the larger the dynamic range and the higher the accuracy of the assay.
6. For optical analysis and to test if Dicer can actually cleave the fluorescent beacon, measure Dicer-mediated fluorescence increase over a certain course of time (Fig. 5). Use UV/visible spectrophotometer with ex = 491 nm and a path length of 10 mm (depends on the sensitivity of your spectrophotometer). Start with pre-miRNA renaturation. Conditions for reaction are as follows: 25 U/ml Dicer, 1× Dicer buffer, and 100 nM pre-miRNA beacon. Measure the fluorescence every 30 min for 4–6 h.
7. Instead of commercial recombinant Dicer it is possible to use 10 % HEK293 cell lysate as a substitute. Especially for Dicer activity testing, cell lysate from any other cell line or tissue may be used. In such settings, we did not observe significant effects upon the addition of RNAse inhibitor.
8. We used the POLAR star Optima (BMG labtech). Any other plate reader may be used, but you might have to adjust the gain factor.

9. RNase A cleaves single-stranded RNA, which leads to unspecific cleavage of the pre-miRNA beacon that could also induce fluorescent increase. To avoid false-positive results, test your assay setup by including 0.3–0.8 U/μl RNase inhibitor in the reaction mixture. The kinetics of fluorescence increase should be independent of the presence of RNAse inhibitor.

Acknowledgement

This work was supported by the German research foundation (DFG).

References

1. Arenz C (2006) MicroRNAs–future drug targets? Angew Chem Int Ed Engl 45(31): 5048–5050
2. Deiters A (2010) Small molecule modifiers of the microRNA and RNA interference pathway. AAPS J 12(1):51–60. doi:10.1208/s12248-009-9159-3
3. Provost P, Dishart D, Doucet J, Frendewey D, Samuelsson B, Radmark O (2002) Ribonuclease activity and RNA binding of recombinant human Dicer. EMBO J 21(21): 5864–5874
4. Bose D, Jayaraj G, Suryawanshi H, Agarwala P, Pore SK, Banerjee R, Maiti S (2012) The tuberculosis drug streptomycin as a potential cancer therapeutic: inhibition of miR-21 function by directly targeting its precursor. Angewandte Chemie 51(4):1019–1023. doi:10.1002/anie.201106455
5. Maiti M, Nauwelaerts K, Herdewijn P (2012) Pre-microRNA binding aminoglycosides and antitumor drugs as inhibitors of Dicer catalyzed microRNA processing. Bioorg Med Chem Lett 22(4):1709–1711. doi:10.1016/j.bmcl.2011.12.103
6. Davies BP, Arenz C (2008) A fluorescence probe for assaying micro RNA maturation. Bioorg Med Chem 16(1):49–55
7. Davies BP, Arenz C (2006) A homogenous assay for micro RNA maturation. Angew Chem Int Ed Engl 45(33):5550–5552
8. Macrae IJ, Zhou K, Li F, Repic A, Brooks AN, Cande WZ, Adams PD, Doudna JA (2006) Structural basis for double-stranded RNA processing by Dicer. Science 311(5758):195–198
9. Vermeulen A, Behlen L, Reynolds A, Wolfson A, Marshall WS, Karpilow J, Khvorova A (2005) The contributions of dsRNA structure to Dicer specificity and efficiency. RNA 11(5):674–682
10. Neubacher S, Dojahn CM, Arenz C (2011) A rapid assay for miRNA maturation by using unmodified pre-miRNA. Chem BioChem 12(15):2302–2305
11. DiNitto JP, Wang LY, Wu JC (2010) Continuous fluorescence-based method for assessing Dicer cleavage efficiency reveals 3' overhang nucleotide preference. Biotechniques 48(4):303–309
12. Mayr F, Schutz A, Doge N, Heinemann U (2012) The Lin28 cold-shock domain remodels pre-let-7 microRNA. Nucleic Acids Res. doi:10.1093/nar/gks355
13. Sambrook J, Russell DW (2001) Molecular cloning: a laboratory manual, 3rd edn. Cold Spring Harbor Laboratory Press, Cold Spring Harbor, NY

Chapter 8

A Fluorescence Correlation Spectroscopy-Based Enzyme Assay for Human Dicer

Eileen Magbanua and Ulrich Hahn

Abstract

We used fluorescence correlation spectroscopy (FCS) to establish an in vitro assay to investigate RNase activity of human Dicer (Werner et al., Biol Chem 393(3):187–193). FCS allows investigating the interactions of different particles due to their differing diffusion mobility, provided that one of the interacting partners exhibit a fluorescence label. In our case we used a fluorophore-labeled double-stranded RNA (dsRNA) as substrate to monitor Dicer activity. The dsRNA was cleaved by the enzyme resulting in a five-nucleotide-short single-stranded RNA (ssRNA) fragment carrying the fluorophore, which could be distinguished from the substrate and unlabeled second product by FCS. Furthermore, we refer to additional (control) experiments to confirm obtained data.

Key words FCS, Human Dicer, Dicer assay, RNase III, dsRNA

1 Introduction

Fluorescence correlation spectroscopy (FCS) is a method, which enables analyzing fluctuations of diffusing fluorescent molecules. Different fluorescent molecules which vary in mass by a factor of eight can be distinguished from each other [1]. Fulfilling this requirement FCS can easily be used to investigate interactions of fluorescently labeled particles and for example in vitro enzyme activity assays [2–4]. Further advantages of FCS are the small sample amounts as well as a small detection volume of approximately 0.1 fl, which represents less than the volume of a standard bacterial cell. Here we used FCS to investigate the RNase activity of human Dicer [5]. Dicer plays a key role during gene regulation by RNA interference (RNAi). It is responsible for the delivery of substrate for RISC by cleaving double-stranded RNA (dsRNA) to small interfering RNAs (siRNA) or microRNAs (miRNAs) which are than interacting with further proteins to target selected mRNAs resulting in gene silencing [6–8].

Christoph Arenz (ed.), *miRNA Maturation: Methods and Protocols*, Methods in Molecular Biology, vol. 1095,
DOI 10.1007/978-1-62703-703-7_8, © Springer Science+Business Media New York 2014

In this assay we used a part of the coding region of glyceraldehyde-3-phosphate dehydrogenase (GAPDH) as Dicer substrate. The substrate was built up of an unlabeled sense strand, which was hybridized to an antisense strand fluorescently labeled at its 5′ end (Fig. 1a). To facilitate optimal substrate binding as well as Dicer cleavage we designed a dsRNA containing a typical two-nucleotide 3′ overhang [9, 10]. The substrate was cleaved into three smaller products: a non-fluorescent 21-nucleotide dsRNA, a uridine 5′ phosphate, and a fluorescing five-nucleotide-short single-stranded RNA (ssRNA) (Fig. 1b). The plotting of resulting autocorrelation curves of substrate and cleavage product demonstrates differences in diffusion time (Fig. 2).

In addition to the assay outlined several control experiments can be performed. The usage of two different fluorophores verifies

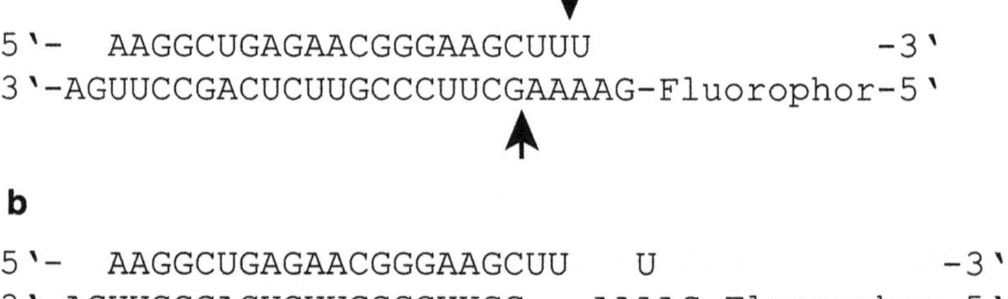

Fig. 1 Dicer substrate. (**a**) dsRNA Dicer substrate deduced from the coding region of GAPDH mRNA with a two-nucleotide protruding 3′ end. Antisense RNA strand possesses a fluorescence label at its 5′-end; sense strand is not labeled. *Arrows* indicate putative cleavage sites. (**b**) Cleavage products after incubation with Dicer are a non-fluorescent 21-nucleotide dsRNA, a uridine 5′ phosphate, and a fluorescing five-nucleotide-long ssRNA

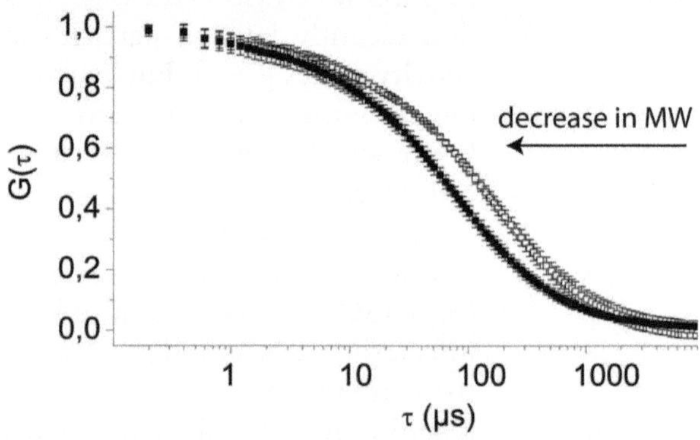

Fig. 2 Autocorrelation functions. Experimentally determined autocorrelation functions of Dicer substrate labeled with ATTO647N (*open boxes*) and cleavage product (*filled boxes*). *Arrow* indicates that components with lower molecular weight (MW) show autocorrelation function curves shifted to the *left*

Dicer activity and reduces the risk of artifacts due to unspecific interactions or photodynamic processes. As Dicer activity is strongly dependent on the presence of Mg^{2+}-ions for instance replacement of Mg^{2+}- by Ca^{2+}-ions leads to a significant inhibition of Dicer activity [11], and thus diffusion time should remain constant due to remaining uncleaved substrate even after enzyme addition. Furthermore, Dicer does solely cleave dsRNA [9]. Therefore fluorescently labeled ssRNA can be used as control as well.

Bacterial RNase III generates Dicer-like termini and overhangs after cleavage of dsRNA, but cleavage products differ in size. RNase III of *Escherichia coli* cleaves dsRNA into 12–15-nucleotide-long RNA fragments. Using GAPDH-derived dsRNA substrate (Fig. 1a) RNase III produces RNA fragments that differ in size from Dicer products. Significant differences should be observed comparing autocorrelation curves as well as diffusion coefficients of cleavage products derived of both enzymes.

2 Materials

Prepare all solutions RNase free (*see* **Note 1**).

2.1 Cleavage Assay

1. Recombinant human Dicer (Genlantis).
2. *E. coli* RNase III (e.g., New England Biolabs).
3. Dicer substrates:

   Sense RNA (5′-AAGGCUGAGAACGGGAAGCUUU-3′).

   Antisense RNA 5′-ATTO647N (5′-ATTO647N-AGUUCC GACUCUUGCCCUUCGAAAAG-3′) (*see* **Note 2**).

   Antisense RNA 5′-Cy5 (5′-Cy5-AGUUCCGACUCUUGCC CUUCGAAAAG-3′).
4. Buffer A: 30 mM Tris–HCl, pH 6.8, 50 mM NaCl.
5. Buffer B: 20 mM Tris–HCl, pH 8.0, 250 mM NaCl.
6. Buffer C: 10 mM Tris–HCl, pH 7.5, 350 mM KCl, 0.1 mM EDTA, 0.1 mM DTT, 2.5 mM $MgCl_2$.
7. 100 mM $MgCl_2$ stock solution.
8. 100 mM $CaCl_2$ stock solution.
9. RNase-free water.

2.2 FCS Measurement

1. Confocal laser scanning microscope (cLSM) with photon counting unit (e.g., ConforCor2 (Carl Zeiss) with software package) (*see* **Note 3**).
2. Cover slip 24 × 50 mm (CarlRoth) (*see* **Note 4**).

3 Methods

3.1 RNA Denaturation and Renaturation

1. Dilute ATTO647N-labeled RNA in Buffer A and combine it with sense RNA equimolecular.
2. Dilute Cy5-labeled RNA in Buffer B and combine it with sense RNA equimolecular.
3. Heat RNA for 3 min at 95 °C.
4. Cool sample down by 1 °C steps per 1.2 min until a temperature of 25 °C is reached.
5. Add $MgCl_2$ to a final concentration of 2.5 mM (*see* **Note 5**).

3.2 FCS Configuration for ATTO647N and Cy5 (See Note 6)

1. Excite with 3 % of 5.0 mW of a HeNe laser at 633 nm.
2. Use the water immersion objective C-Apochromat 40× with numerical aperture of 1.2.
3. Separate emitted light from detection volume with a 633 nm main dichromic mirror (HFT) and 650 nm emission long-pass filter.
4. Adjust pinhole to 90 μm.
5. Measurement period: 5 or 10 s.

3.3 FCS Measurements

Measurements were carried out in a sample volume of 20 μl positioned on cover slide in cLSM at room temperature.

1. Measure diffusion time of 5 nM fluorescent dsRNA in the absence of enzyme in appropriate buffer.
2. Add Dicer to a final concentration of 1 U directly to the substrate on cover slide, and measure diffusion time again. Diffusion time decreases significantly due to generation of cleavage product (*see* **Note 7**).
3. Plot autocorrelation curves of substrate and Dicer product against diffusing time and compare (Fig. 2). The smaller the molecular weight of a diffusing component the more the autocorrelation curve is shifted to the left.
4. Optional control experiment 1: Perform Dicer assay with another Dicer substrate, which is labeled with an alternative fluorophore.
5. Optional control experiment 2: After RNA denaturation and renaturation add $CaCl_2$ to a final concentration of 2.5 mM instead of $MgCl_2$ to RNA (*see* **Note 8**). Then perform Dicer assay as described in **steps 1–3**. No reduction of diffusion time should be observed.
6. Optional control experiment 3: Perform Dicer assay with single-stranded fluorescent RNA as substrate. No reduction regarding diffusion time should be observed (*see* **Note 9**).

7. Optional control experiment 4:
   Use Cy5-labeled substrate and perform Dicer assay with Dicer enzyme and RNase III in parallel. In case of RNase III use 2.6 U enzyme and buffer C.

   Subsequently, compare autocorrelation curves of substrate alone with products after Dicer and RNase III cleavage, respectively. Autocorrelation curve should differ significantly due to unequal cleavage products.

   Define diffusion time τ of cleavage products by using the two-component model. Therefore use instrument software and insert diffusion time of uncleaved substrate as fixed value. The calculated value represents diffusion time of respective cleavage products.

   Finally, calculate diffusion coefficients D of cleavage products with

$$D = \frac{r_0^2}{4 \cdot \tau}$$

   and use for the radial distance to the center of the laser beam focus $r_{0,\mathrm{Cy5}} = 250$ nm and diffusion time τ of cleavage product, respectively. Determined diffusion coefficients should differ significantly.

4 Notes

1. For RNase-free water specific filter installation can be used. Optionally solutions can be treated with diethylpyrocarbonate (DEPC) to inactivate RNases due to acetylation of histidine, lysine, cysteine, and tyrosine residues. Therefore solution is treated with 0.1 % (v/v) DEPC and stirred overnight. Autoclaving afterwards leads to destruction of DEPC.
2. Of course RNA can be labeled with two other fluorophores. Ideally both fluorophores should possess same absorption and emission spectra and be suitable for FCS measurements.
3. Using a software package facilitates the usage of FCS as well as analysis of data sets to apply different component models (e.g., two-component model) to calculate diffusion time of a second diffusion compound for instance. Otherwise all parameters would have to be calculated in the old-fashioned way by hand.
4. Thickness of cover slip has to enable performing confocal measurements.
5. Heating RNA in the presence of $MgCl_2$ might lead to hydrolysis of RNA.
6. Using other fluorophores these parameters have to be adjusted.

7. Investigating two different components by FCS, both have to differ significantly regarding diffusion time or molar mass, respectively.
8. Replacing divalent cations leads to inhibition of Dicer activity.
9. Dicer exclusively cleaves dsRNA and not ssRNA. Nevertheless, diffusion time can increase slightly after Dicer addition due to Dicer affinity to 5′-terminal modifications of ssRNA [12].

References

1. Weisshart K, Jungel V, Briddon SJ (2004) The LSM 510 META - ConfoCor 2 system: an integrated imaging and spectroscopic platform for single-molecule detection. Curr Pharm Biotechnol 5(2):135–154
2. Kinjo M, Nishimura G, Koyama T et al (1998) Single-molecule analysis of restriction DNA fragments using fluorescence correlation spectroscopy. Anal Biochem 260(2):166–172
3. Ketting RF, Fischer SE, Bernstein E et al (2001) Dicer functions in RNA interference and in synthesis of small RNA involved in developmental timing in *C. elegans*. Genes Dev 15(20):2654–2659
4. Kohl T, Haustein E, Schwille P (2005) Determining protease activity in vivo by fluorescence cross-correlation analysis. Biophys J 89(4):2770–2782
5. Werner A, Skakun VV, Ziegelmuller P et al (2012) A fluorescence correlation spectroscopy-based enzyme assay for human Dicer. Biol Chem 393(3):187–193
6. Fire A, Xu S, Montgomery MK et al (1998) Potent and specific genetic interference by double-stranded RNA in *Caenorhabditis elegans*. Nature 391(6669):806–811
7. Jinek M, Doudna JA (2009) A three-dimensional view of the molecular machinery of RNA interference. Nature 457(7228):405–412
8. Pillai RS, Bhattacharyya SN, Filipowicz W (2007) Repression of protein synthesis by miRNAs: how many mechanisms? Trends Cell Biol 17(3):118–126
9. Zhang H, Kolb FA, Brondani V et al (2002) Human Dicer preferentially cleaves dsRNAs at their termini without a requirement for ATP. EMBO J 21(21):5875–5885
10. Vermeulen A, Behlen L, Reynolds A et al (2005) The contributions of dsRNA structure to Dicer specificity and efficiency. RNA 11(5): 674–682
11. Provost P, Dishart D, Doucet J et al (2002) Ribonuclease activity and RNA binding of recombinant human Dicer. EMBO J 21(21): 5864–5874
12. Kini HK, Walton SP (2007) In vitro binding of single-stranded RNA by human Dicer. FEBS Lett 581(29):5611–5616

Chapter 9

Detection of microRNA Maturation Using Unmodified pre-microRNA and Branched Rolling Circle Amplification

Saskia Neubacher and Christoph Arenz

Abstract

The ever-increasing number of different miRNAs and their association with a vast number of cellular dysfunctions and diseases have initiated several groups to investigate miRNA maturation, which ultimately leads to down regulation of a target messenger RNA (mRNA) and its downstream product. A rapid, convenient, and reliable assay to detect the Dicer-mediated miRNA-maturation step may facilitate research in this field. Here we describe the in vitro detection of the Dicer-mediated miRNA maturation step using unmodified pre-miRNA and branched rolling circle amplification.

Key words miRNA maturation assay, Enzyme assay, Rolling circle amplification, Primer extension, High-throughput screening

1 Introduction

The ever-increasing number of different microRNAs (miRNAs) and their association with a vast number of cellular dysfunctions and diseases have initiated several groups to investigate miRNA maturation, which ultimately leads to down regulation of a target messenger RNA (mRNA) and its downstream product [1–5]. miRNAs are approximately 22-nucleotide (nt) double-stranded (ds) RNAs and are generated in two consecutive enzymatic steps from the primary miRNA (pri-miRNA) that is transcribed by RNA polymerase II. After the pri-miRNA is cleaved in the nucleus by the endoribonuclease Drosha, the resulting precursor miRNA (pre-miRNA) is exported into the cytosol by exportin-5. There, the endoribonuclease Dicer cleaves its substrate independent of its sequence and rather acts like a molecular ruler. Dicer recognizes the end of the pre-miRNA stem with its 5′-OH and 2-nt 3′-overhang and then cleaves two phosphodiester bonds about 21–23 nt into the stem also creating this signature 3′- and 5′-end [6–10].

The main method to detect Dicer-mediated cleavage of pre-miRNAs to mature miRNAs has been through northern blotting

Christoph Arenz (ed.), *miRNA Maturation: Methods and Protocols*, Methods in Molecular Biology, vol. 1095, DOI 10.1007/978-1-62703-703-7_9, © Springer Science+Business Media New York 2014

[11]. This method involves radioactive labeling of RNA and tedious polyacrylamide gel electrophoresis, making it unsuitable for high-throughput screening. Qiagen offers a quantitative PCR kit for the separate quantification of miRNA and pre-miRNA from one sample [12], and researchers from Pfizer have published an assay to evaluate Dicer cleavage efficiency of double-stranded RNA [13]. The first in vitro assay that directly monitors Dicer-meditated miRNA maturation has been developed in our laboratory and uses fluorecently labeled pre-miRNAs. Therein, the 5′-end of the pre-miRNA is attached to a fluorophore and the 3′-end to a quencher, which leads to absorption of the emitted fluorescence presumably through collisional quenching when the typical hairpin structure is formed. Incubation with Dicer produces small double-stranded miRNAs and leads to ineffective quenching and increase in fluorescence, and this can then be used to monitor the Dicer cleavage process [14, 15]. However, as mentioned above, it is not clear how these pre-miRNA modifications influence the Dicer cleavage. It has been reported that modifications to the 5′- and 3′-ends can cause a shift in the cleavage site and, hence, a modified mature miRNA sequence, and that the efficiency and specificity of the cleavage through Dicer are influenced [10]. Therefore, modifications of the pre-miRNA for a miRNA maturation assay should be avoided. Though reporter gene assays exist and have been used to screen small-molecule libraries for the inhibition of miRNA maturation, the exact molecular target of thus-identified inhibitors remains unclear and has to be examined in further experiments [16, 17]. The group of Maiti has circumvented this limitation by screening a number of amingoglycosides—known as RNA binders—with such a reporter gene assay [18]. The inhibition of pre-miR-21 maturation was afterwards confirmed by a maturation assay using ^{32}P-labeled RNA. An in vitro assay allowing the detection of solely the Dicer-mediated cleavage of pre-miRNA to mature miRNA would allow an immediate evaluation of test compounds.

Jonstrup et al. [19] and Cheng et al. [20] applied the known isothermal branched rolling circle amplification (BRCA) technique to quantify miRNAs in cell lysate. To enhance specificity in their quantification method, they use a ligase for the miRNA-templated circularization of the ssDNA padlock probe. However, both groups did not clarify the issue of selectivity between pre-miRNA and mature miRNA. We have modified this detection method towards a rapid functional in vitro assay to detect miRNA maturation using unmodified pre-miRNA. Therein, a designed circular ssDNA (*see* **Note 1**) carries a hybridization site for the mature miRNA (Fig. 1). The miRNA that originates from the 5′-arm of the pre-miRNA then serves as a primer for isothermal amplification with *Bacillus stearothermophilus* (Bst) DNA poylmerase (large fragment, *see* **Note 2**). The usage of a secondary primer complementary to the generated concatamer enhances amplification. Unlike the mature miRNA, the pre-miRNA cannot serve as a primer, since the 3′-OH

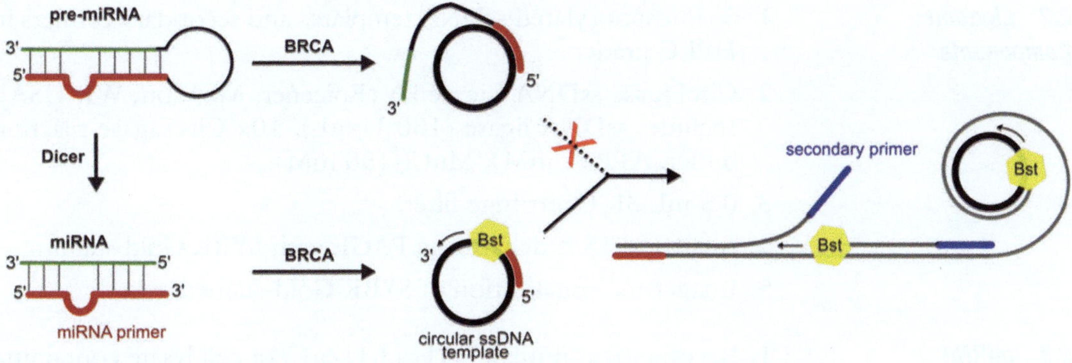

Fig. 1 Schematic representation of the BRCA-based miRNA maturation assay [5]

adjacent to the miRNA hybridization site is blocked (Fig. 1, top). The usage of SYBR Gold during the amplification enables the detection of amplicon formation in real time, and a comparison of the resulting slopes of fluorescence increase over time to control measurements allows an evaluation of test compounds towards the inhibition of Dicer-mediated miRNA maturation or of different Dicer-containing probes [5].

2 Materials

All materials used for RNA handling were ordered pyrogen, DNase, and RNase free, if possible. Glass- and metalware were heated to 250 °C for 3 h. All surfaces were regularly treated with an aqueous sodium hypochlorite solution (approx. 3 %) and then rinsed with ultrapure water. Sterile gloves were used when handling RNA-containing solutions. All buffers were filtered using a 0.22-µm sterile filter.

2.1 In Vitro Transcription Components

1. T7 ssDNA template and corresponding T7 complementary sequences for in vitro transcription of pre-miRNAs in HPLC grade.
2. In vitro transcription kit (data optimized with T7 RiboMAX Expression Large Scale RNA Production System, P1300, Promega, Madison, WI, USA).
3. Triethylammoniumacetate (TEAA) buffer (0.1 M, pH 7.0) and acetonitrile for HPLC purification.
4. Reverse-phase HPLC column, e.g., Varian Polaris C18 A 5 µm 250 × 046 (pore size: 200 Å), heated to 55 °C.
5. miRNA Marker and low-range ssRNA marker.
6. Analytical 15 % denaturing PAGE with SYBR Gold staining.
7. Imager for visualization of SYBR Gold-stained gels.

2.2 Ligation Components

1. 5′-Phosphorylated ssDNA templates and secondary primers in HPLC grade.
2. CircLigase ssDNA Ligase kit (Epicener, Madison, WI, USA). Includes ssDNA ligase (100 U/μL), 10× CircLigase reaction buffer, ATP (1 mM), $MnCl_2$ (50 mM).
3. 0.5 mL 3K Centrifuge filter.
4. Analytical 15 % denaturing PAGE with SYBR Gold staining.
5. Imager for visualization of SYBR Gold-stained gels.

2.3 miRNA Maturation

1. Recombinant human Dicer (1 U/μL) or cell lysate containing RNAse inhibitor.
2. Pre-miRNA.
3. Dicer buffer: 20 mM Tris–HCl (pH 6.8), 12.5 mM NaCl, 2.5 mM $MgCl_2$, 1 mM DTT.
4. Dicer dilution buffer: 20 mM Tris–HCl (pH 6.8), 100 mM NaCl, 1 mM $MgCl_2$, 1 mM DTT, 5 mM β-mercaptoethanol, 10 % glycerol, 0.1 % Triton-X 100.
5. Clear 96-well plates.
6. Incubator (*see* **Note 3**).

2.4 BRCA Components

1. Circular ssDNA and secondary primer.
2. Dicer-processed pre-miRNA.
3. Bst polymerase (8 U/μL; New England Biolabs), including ThermoPol buffer.
4. Deoxynucleoside triphosphates (dNTP-mix).
5. Clear 96-well plates.
6. Real-time thermocycler (*see* **Note 4**).

3 Methods

3.1 In Vitro Transcription

1. T7 ssDNA template and corresponding T7 complementary sequences for in vitro transcription of pre-miRNAs were designed (*see* **Note 5**).
2. Before the addition of T7 RNA polymerase, heat the T7 ssDNA template and T7 primer in transcription buffer to 90 °C and snap cool on ice to ensure correct hybridization. In vitro transcription is essentially done according to the manufacturer's protocol in a 100-μL scale. Extend the incubation time to 2–4 h. After treatment with DNase for 30 min at 37 °C (*see* **Note 6**), dilute the RNA solution with H_2O to a volume of approximately 100–200 μL. Then extract with same volume phenol/chloroform/isoamyl alcohol (PCI, 25:24:1), and then

precipitate with 90 % ethanol and 0.5 M NH_4OAc by incubation at −80 °C overnight.

3. Wash precipitated RNA with ice-cold 85 % ethanol and dissolve in water (approx. 500 μL). After filtration through a 0.22-μm sterile microfilter, load the RNA onto a semi-preparative reversed-phase HPLC column. Run a linear gradient increasing the acetonitrile fraction from 3 to 30 % in 40 min. Optimal gradients should be determined experimentally. Fractions containing RNA product should immediately be sealed and cooled on ice and subsequently be analyzed by MALDI-TOF or denaturing PAGE.
4. The correct fractions are lyophilized or again precipitated from ethanol. Dissolve RNA in H_2O, and determine the concentration using the UV absorption at $\lambda = 260$ nm and the calculated molar extinction coefficient (nearest neighbor) [21].
5. Analyze the transcribed pre-miRNA by MALDI-TOF and 15 % denaturing PAGE using the miRNA marker and low-range ssRNA marker.

3.2 Circularization of 5′-Phosphorylated ssDNA

1. 5′-Phosphorylated ssDNA templates and secondary primers were designed according to the pre-miRNA used (*see* **Note 7** and **8**).
2. For a 40 μL reaction, mix H_2O (26 μL), CircLigase buffer (4 μL, 10×), ATP (2 μL, 1 mM), $MnCl_2$ (2 μL, 50 mM), and 5′-phosphorylated ssDNA (4 μL, 10 μM). Then add CircLigase ssDNA ligase (2 μL, 100 U/μL) and mix gently. Incubate the reaction mixture at 60 °C for 4 h and then at 16 °C for 12 h. Inactivate the enzyme by heating to 80 °C for 10 min.
3. Dilute the ligation mixture to 300 μL with Tris–HCl buffer (10 mM, pH 8). Transfer the solution to an Amicon Ultra-0.5 mL 3K centrifuge filter and centrifuge (12,000×*g*, 20 min, RT). Discard the filtrate, add another 300 μL buffer, and centrifuge again. Add 30 μL buffer to elute the circular ssDNA template.
4. Calculate the concentration by using the calculated molar extinction coefficient (nearest neighbor method) [21] and determining the absorption at $\lambda = 260$ nm (A_{260}).
5. For quality control analyze the ligation product (approx. 1 pmol) with a 15 % denaturing PAGE and SYBR Gold staining. The circular ssDNA template should appear at higher molecular weights relative to the linear substrate (Fig. 2).

3.3 miRNA Maturation

1. In a 200-μL tube, prepare a 1 μM pre-miRNA stock solution in Dicer buffer. Place the tube into a thermocycler and heat to 90 °C for 5 min. Then slowly cool to 25 °C at 2 °C/min. This stock solution can be stored at −20 °C for up to 2 weeks (*see* **Note 9**).

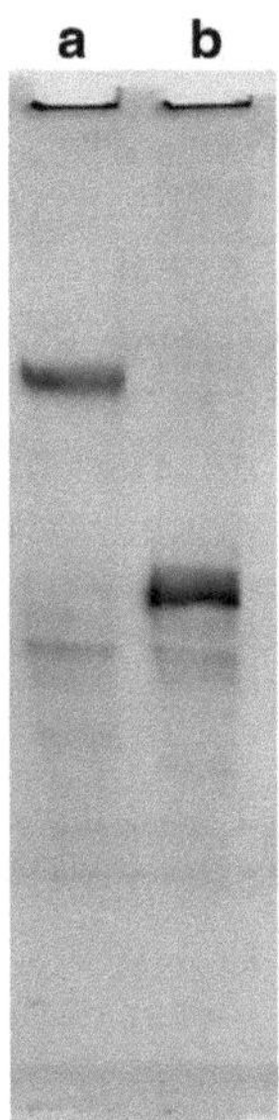

Fig. 2 15 % Denaturing PAGE of (**a**) circular ssDNA template and (**b**) linear 5′-phosphorylated ssDNA (SYBR Gold stain, $\lambda_{ex} = 470$ nm, $\lambda_{em} = 535$ nm)

2. Dilute Dicer (1 U/μL) with Dicer dilution buffer to 0.04 U/μL.
3. Use an aliquot of diluted Dicer to prepare heat-denatured Dicer for control reactions. Heat to 90 °C for 15 min, and store at −20 °C.
4. In a clear 96-well plate, 20-μL reactions were performed in at least three replicates per test compound (*see* **Notes 10** and **11**). Calculate the volume needed for all reactions, and prepare a pre-miRNA dilution (71.4 nM) in Dicer buffer. For ten reactions, for example, dilute pre-miRNA (11 μL, 1 μM) in Dicer buffer (143 μL, 1×).
5. Add the test compound in the desired concentration or solvent (4 μL) and pre-miRNA (14 μL) into each well. Mix by repeated aspiration, and incubate at RT for 30 min.
6. Add Dicer (2 μL) to each well and mix by repeated aspiration. Incubate at 37 °C for 2 h in a thermocycler with heated lid. Increase temperature to 90 °C for 15 min for inactivation, and store on ice.

3.4 Rolling Circle Amplification

1. Towards the end of the Dicer incubation, prepare a master-mix solution for the BRCA reaction on ice. Again consider all reactions including all controls. The volume that is added to each BRCA reaction should be 10 μL, including 2× ThermoPol buffer, 0.4 mM dNTPs, 0.4 μM secondary primer, 1:10,000 (v/v) SYBR Gold, 10 % DMSO, and 4 nM circular ssDNA template.

2. Preheat the real-time thermocycler to 57 °C.
3. Transfer 5 μL of each Dicer reaction into a pre-cooled, clear 96-well plate on ice. Add 4 μL H_2O to controls and 4 μL test compound in the appropriate concentrations to all other reactions.
4. First add 10 μL BRCA master-mix solution, and then add 1 μL Bst DNA polymerase to each well, mixing thoroughly after each addition by repeated aspiration. Avoid the formation of bubbles, as these can disturb fluorescence measurements.
5. Close the well plate, and place it into the preheated thermocycler. Measure in 2-min increments at 57 °C and $\lambda_{ex} = 485 \pm 20$ nm and $\lambda_{em} = 530 \pm 20$ nm for 2 h.
6. Calculate the initial linear slopes followed by IC_{50} values (*see* **Note 12** and Fig. 3).

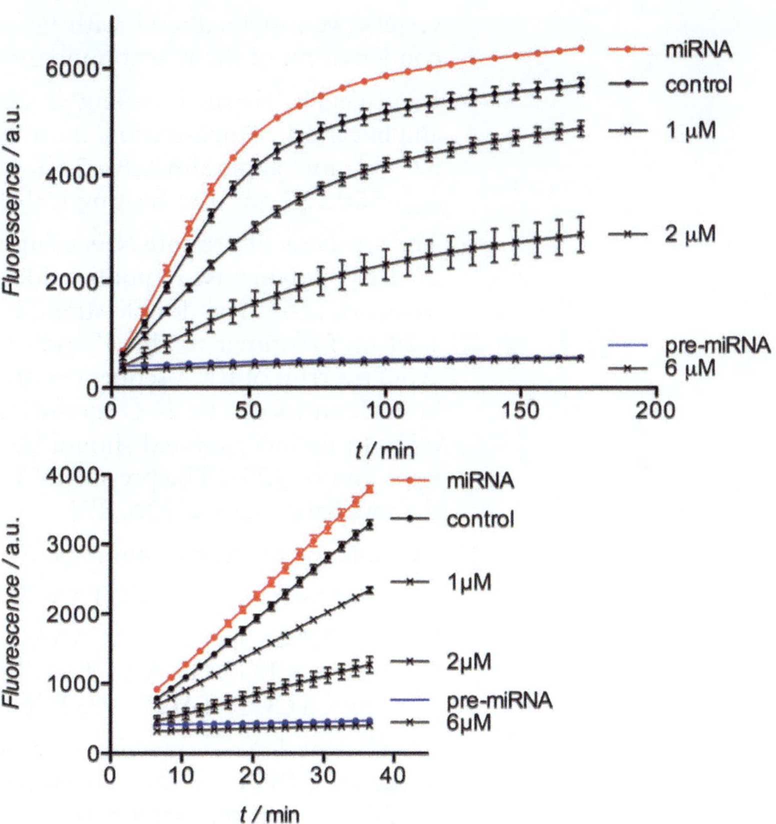

Fig. 3 Data of a BRCA-based miRNA maturation assay. miRNA maturation without test compound shows the greatest slope (0 % inhibition), and miRNA maturation with denatured Dicer shows no fluorescence increase (100 % inhibition). The control where the test compound is added after Dicer incubation shows only a minor decrease in fluorescence increase. Error bars represent the standard deviation of three replicates [5]

4 Notes

1. The miRNA assay as described here directly uses circular ssDNA omitting the ligation of the padlock probe, as the focus is not the specific quantification of very low concentrations of a specific miRNA amongst other miRNAs but rather the detection of the maturation process of the miRNA. Should there be specificity problems, it is possible to use a linear ssDNA as a padlock probe with a miRNA-dependent ligation step before isothermal amplification [20, 22, 23].
2. Important to note is that the DNA polymerase must lack exonuclease activities, specially in 3′–5′ direction, as this can lead to false generation of the mature miRNA that is used as the primer to initiate the BRCA. The amplification signal would then not be Dicer dependent.
3. The incubator should guarantee heating of the whole reaction vessel (oven or incubator with heated lids) in order to avoid condensation of the reaction mixture.
4. In principle, thermal cycling is not needed for isothermal amplification. Amplification must be monitored at constant 57 °C for approximately 2 h at $\lambda_{ex} = 485 \pm 20$ nm and $\lambda_{em} = 530 \pm 20$ nm. Lid-heating is therefore essential.
5. The synthesis of pre-miRNAs using in vitro transcription with T7 RNA polymerase requires a double-stranded T7 promoter sequence [24]. The double strand is generated by hybridizing a 20-nt T7 primer to the 3′-end of the ssDNA template. To avoid overrun of RNA sequences, the two 5′-terminal deoxyribonucleotides can be 2′-*O*-methylated. Ideally, the first nucleotide to be incorporated should be a GTP to ensure efficient transcription [25]. The pre-miRNA sequences can be found in the miRBase database [26, 27].

   Example for *Drosophila melanogaster* let-7:

   T7 primer sequence: 5′-GGTAATACGACTCACTATAG-3′.

   let-7 ssDNA template: 5′-ACAAAGCTAGCACATTGTATAG TATGATGTGTAATTACTACTATACAACCTAC TACCTCCTATAGTGAGTCGTATTA-3′.
6. An efficient DNAse reaction is crucial for efficient miRNA-triggered BRCA. If low fluorescence increase is observed, troubleshooting may start here.
7. It should be noted that the 5′-arm of the pre-miRNA always serves as the miRNA primer and never the 3′-arm. The mature miRNA is double stranded and includes the guide and the passenger strand. Depending on the miRNA and the tissue it is expressed in both or either can have a target mRNA. Therefore, the primer can but does not always have to be equal to the mature guide miRNA.

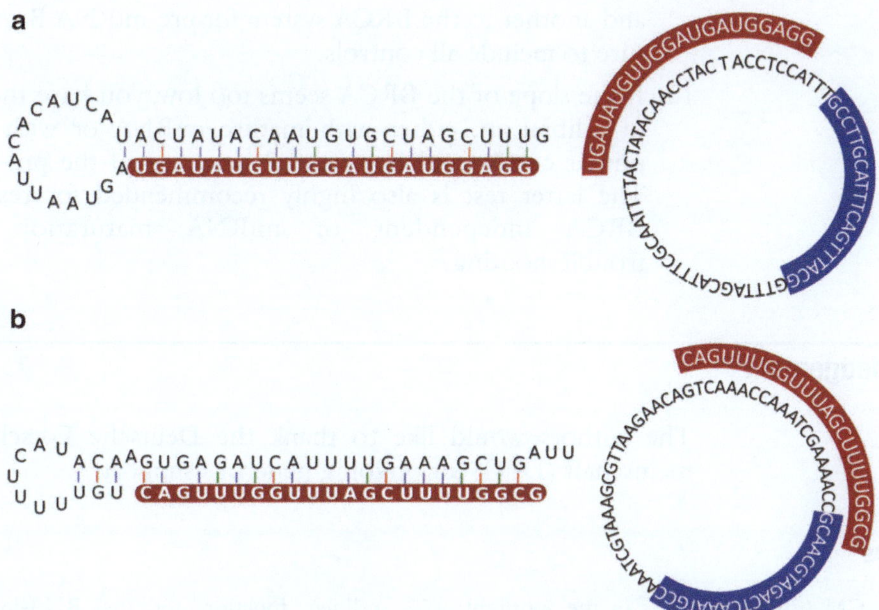

Fig. 4 Design and sequences of a BRCA system for *Drosophila melanogaster* pre-let-7/let-7 (**a**) and pre-bantam/bantam (**b**). The mature miRNA strand used for amplification is shown in *red*, and the secondary primer is shown in *blue*

8. In principle, the basic rules of standard PCR apply to the design of the oligonucleotides for BRCA. The 5′-phosphorylated ssDNA sequences for circular ssDNA templates include a hybridization sequence for the miRNA and for the secondary primer. Self-hybridization and dimer formation of the ssDNA template and the secondary primer should be avoided. The circular ssDNA should be around 65 nt long [12]. Examples for the pre-miRNA and the corresponding circular ssDNA are shown for *D. melanogaster* pre-let-7 and pre-bantam in Fig. 4 [5].
9. Pre-miRNA is best stored as a pellet after standard ethanol precipitation (85 % ethanol in 0.5 M NH_4OAc (pH 7.5) at −80 °C).
10. Be sure to include all controls for the experiment: (1) pre-miRNA with denatured Dicer and without test compound; (2) pre-miRNA with Dicer and without test compound; and (3) pre-miRNA with Dicer and test compound added before BRCA, but after Dicer maturation.
11. The miRNA maturation assay can also be performed simultaneously for more than one pre-miRNA for example to determine selectivities of test compounds. In that case, the miRNA maturation is performed in the presence of pre-miRNA A and B (both at 50 nM final concentrations). An aliquot is then added to the corresponding BRCA system for pre-miRNA A

and another to the BRCA system for pre-miRNA B. Again, be sure to include all controls.

12. If the slope of the BRCA seems too low, you have the chance of calibration, either with mature miRNA or with a DNA primer corresponding to the 5′-segment of the pre-miRNA. The latter test is also highly recommended for testing the BRCA independent of miRNA maturation during troubleshooting.

Acknowledgement

The authors would like to thank the Deutsche Forschungsgemeinschaft (DFG) for ongoing generous support.

References

1. Croce CM (2011) miRNAs in the spotlight: understanding cancer gene dependency. Nat Med 17:935–936
2. Pfeifer A, Lehmann H (2010) Pharmacological potential of RNAi–focus on miRNA. Pharmacol Therapeut 126:217–227
3. Medina PP, Slack FJ (2008) microRNAs and cancer: an overview. Cell Cycle 7:2485–2492
4. Arenz C (2006) MicroRNAs-future drug targets? Angew Chem Int Ed Engl 45: 5048–5050
5. Neubacher S, Dojahn CM, Arenz C (2011) A rapid assay for miRNA maturation by using unmodified pre-miRNA. Chembiochem 12:2302–2305
6. Takeshita D, Zenno S, Lee WC, Nagata K, Saigo K, Tanokura M (2007) Homodimeric structure and double-stranded RNA cleavage activity of the C-terminal RNase III domain of human dicer. J Mol Biol 374:106–120
7. MacRae IJ, Doudna JA (2007) Ribonuclease revisited: structural insights into ribonuclease III family enzymes. Curr Opin Struct Biol 17:138–145
8. Hellani A, Schuchman EH, Al-Odaib A, Al Aqueel A, Jaroudi K, Ozand P, Coskun S (2004) Preimplantation genetic diagnosis for Niemann-Pick disease type B. Prenat Diagn 24:943–948
9. Zhang H, Kolb FA, Jaskiewicz L, Westhof E, Filipowicz W (2004) Single processing center models for human Dicer and bacterial RNase III. Cell 118:57–68
10. Starega-Roslan J, Koscianska E, Kozlowski P, Krzyzosiak WJ (2011) The role of the precursor structure in the biogenesis of microRNA. Cell Mol Life Sci 68:2859–2871
11. Hutvagner G, McLachlan J, Pasquinelli AE, Balint E, Tuschl T, Zamore PD (2001) A cellular function for the RNA-interference enzyme Dicer in the maturation of the let-7 small temporal RNA. Science 293:834–838
12. miScript II RT Kit, Cat. # 218161.
13. DiNitto JP, Wang LY, Wu JC (2010) Continuous fluorescence-based method for assessing Dicer cleavage efficiency reveals 3 ′ overhang nucleotide preference. Biotechniques 48:303–309
14. Davies BP, Arenz C (2006) A homogenous assay for micro RNA maturation. Angew Chem Int Ed Engl 45:5550–5552
15. Davies BP, Arenz C (2008) A fluorescence probe for assaying micro RNA maturation. Bioorg Med Chem 16:49–55
16. Gumireddy K, Young DD, Xiong X, Hogenesch JB, Huang Q, Deiters A (2008) Small-molecule inhibitors of microrna miR-21 function. Angew Chem Int Ed Engl 47:7482–7484
17. Young DD, Connelly CM, Grohmann C, Deiters A (2010) Small molecule modifiers of microRNA miR-122 function for the treatment of hepatitis C virus infection and hepatocellular carcinoma. J Am Chem Soc 132:7976–7981
18. Bose D, Jayaraj G, Suryawanshi H, Agarwala P, Pore SK, Banerjee R, Maiti S (2012) The tuberculosis drug streptomycin as a potential cancer therapeutic: inhibition of miR-21 function by directly targeting its precursor. Angew Chem Int Ed 51:1019–1023
19. Jonstrup SP, Koch J, Kjems J (2006) A microRNA detection system based on padlock probes and rolling circle amplification. RNA 12:1747–1752
20. Cheng Y, Zhang X, Li Z, Jiao X, Wang Y, Zhang Y (2009) Highly sensitive determination of microRNA using target-primed and branched rolling-circle amplification. Angew Chem Int Ed 48:3268–3272

21. For example http://www.ribotask.com/calculator.asp.
22. Neubacher S, Arenz C (2009) Rolling-circle amplification: unshared advantages in miRNA detection. ChemBioChem 10: 1289–1291
23. Harcourt EM, Kool ET (2012) Amplified microRNA detection by templated chemistry. Nucleic Acids Res 40:e65
24. Milligan JF, Groebe DR, Witherell GW, Uhlenbeck OC (1987) Oligoribonucleotide synthesis using T7 RNA polymerase and synthetic DNA templates. Nucleic Acids Res 15:8783–8798
25. Milligan JF, Uhlenbeck OC (1989) Synthesis of small RNAs using T7 RNA polymerase. Methods Enzymol 180:51–62
26. http://www.mirbase.org
27. Kozomara A, Griffiths-Jones S (2011) miRBase: integrating microRNA annotation and deep-sequencing data. Nucleic Acids Res 39:D152–D157

Chapter 10

Quantitative RT-PCR Specific for Precursor and Mature miRNAs

Hannah Zöllner, Stephan A. Hahn, and Abdelouahid Maghnouj

Abstract

Quantification of microRNAs (miRNAs) in cells or primary tissues is one of the most important steps in elucidating their biological functions. However, miRNAs are challenging molecules for PCR amplification due to the stable hairpin of the precursor form and the small size of the mature miRNA, which is roughly the same length as the primers used in standard PCR. To date, different assays were introduced for the specific and sensitive quantification of the mature form of miRNAs. In this chapter we describe the extraction of RNA from microdissected tissue and the quantification of miRNAs using two different methods (stem–loop qRT-PCR and polyT adaptor qRT-PCR).

Key words microRNA, Stem–loop PCR, TaqMan®, polyT adaptor PCR, qRT-PCR

1 Introduction

Accurate information of spatiotemporal expression patterns of miRNAs is a crucial step for understanding their functions, requiring quantitative detection technologies specific for the active (mature) miRNA rather than the inactive pre-miRNAs. Specific detection of the mature miRNA molecules by conventional RNA techniques is challenging, mainly due to their small size, end polymorphisms, heterogeneous GC content, and sequence homology. In order to reach this goal, adaptions of the standard quantitative reverse transcription PCR (qRT-PCR) have been established for the detection of miRNAs and are now widely used. Standard qRT-PCR primer design is incompatible with miRNA detection because the short size of the target molecule prohibits the accommodation of both forward and reverse PCR primers. Two technologies were introduced, the so-called stem–loop qRT-PCR and polyT adaptor qRT-PCR to overcome this limitation [1, 2].

Christoph Arenz (ed.), *miRNA Maturation: Methods and Protocols*, Methods in Molecular Biology, vol. 1095, DOI 10.1007/978-1-62703-703-7_10,

1.1 Principle of Stem–Loop qRT-PCR

The stem–loop oligonucleotide design introduced by Applied Biosystems is able to specifically reverse transcribe only the selected mature miRNA which is subsequently amplified by qRT-PCR (Fig. 1) [3]. A typical stem–loop reverse transcription (RT) primer includes a 5–8 nt 3′ overhang sequence complementary to the 3′ end of the mature miRNA, a stem and a loop sequence which also contains a universal 3′ priming site for the subsequent qRT-PCR step. The idea behind the stem loop design is that base stacking and spatial constraint of the stem–loop structure improves specificity and sensitivity of the assay. Furthermore, it may also prevent binding of the RT primer to double-strand genomic DNA molecules, thus RNA sample preparation can be omitted. Lastly, stem–loop RT primers can be used for multiplex RT reactions. Apart from the requirements of a miRNA-specific stem–loop oligonucleotide for RT, the subsequent PCR steps include only standard qRT-PCR technology. A specific feature of the herein described TaqMan® miRNA qRT-PCR technology from Applied Biosystems is the use of the so-called TaqMan® probe which lends higher specificity to the assay than conventional qRT-PCRs without this probe.

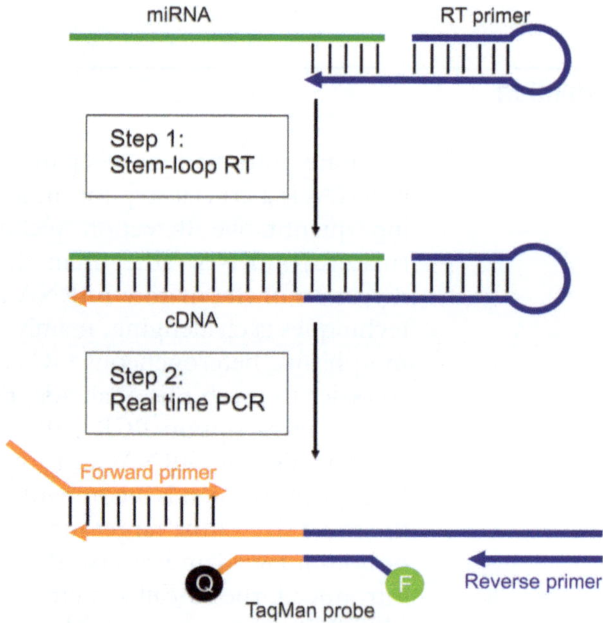

Fig. 1 Schematic description of TaqMan-based real-time quantification of miRNAs. This technique includes two steps: stem–loop RT and real-time PCR. Stem–loop RT primers bind to the 3′ portion of miRNA molecules and are reverse transcribed. Subsequently, the RT product is quantified using conventional TaqMan PCR that includes miRNA-specific forward primer, reverse primer, and a dye-labeled TaqMan probe (figure adapted from ref. 3)

The TaqMan® probes are designed to be complementary to an internal stretch of the amplicon (here they are complementary for the chosen mature miRNA sequence). In addition, these probes contain a fluorophore and a quencher covalently attached to the 5′ and 3′ end of the oligonucleotide probe, respectively. The close proximity of the fluorophore to the quencher molecule prevents the emission of fluorescence. During PCR, the Taq-polymerase extends the primer and reaches the 5′ end of the probe which is hydrolyzed by the 5′→3′ exonuclease activity of this enzyme. As a result, the fluorescent reporter dye is no longer in close proximity to the quenching group and fluorescence increases proportional to the amount of generated PCR products.

1.2 Principle of polyT Adaptor qRT-PCR

Unlike mRNAs, mature miRNAs are not polyadenylated. In this method (Fig. 2), mature miRNAs are elongated by adding a polyA-tail to their 3′ end during the reverse transcription step using *Escherichia coli* PolyA polymerase (PAP). The polyA-tailed miRNAs are then converted into cDNA using a modified oligo-dT primer which also contains a universal tag sequence on its 5′ end for subsequent PCR amplification. Although this strategy may be somewhat less specific than the above-described TaqMan approach, it has nevertheless a number of advantages: (1) Synthesized cDNA can be used for any available compatible miRNA assay and thus helps to safe RNA if it is limited. (2) For detection of precursor miRNA, only an additional specific reverse primer targeting the stem–loop sequence is required (Fig. 2b). (3) This technology uses the rather cheap SYBR Green technology to detect the amplification products.

1.3 Microdissection

The isolation and characterization of homogenous cell populations are of great importance in order to attribute changes in the biological state of a tissue, i.e., the miRNA expression to a specific cell type. In most cases, such an attribution is not possible if bulk tissue is used because the source to prepare the analyte is very heterogenous due to the composition of most tissues containing various different cell types. Several methods have been developed for the isolation of homogenous cell populations from within a mixed cell population; most of them are variations of the so-called laser capture microdissection (LCM). Manual microdissection is a simple and cheap alternative to LCM also able to isolate tissues or cells of interest with sufficient high precision. We and others routinely use manual microdissection for our miRNA expression analyses. In contrast to the LCM, manual microdissection offers greater protection to biomolecules such as DNA, RNA, or proteins [4], likely because no laser energy is required, for what RNA is particularly sensitive.

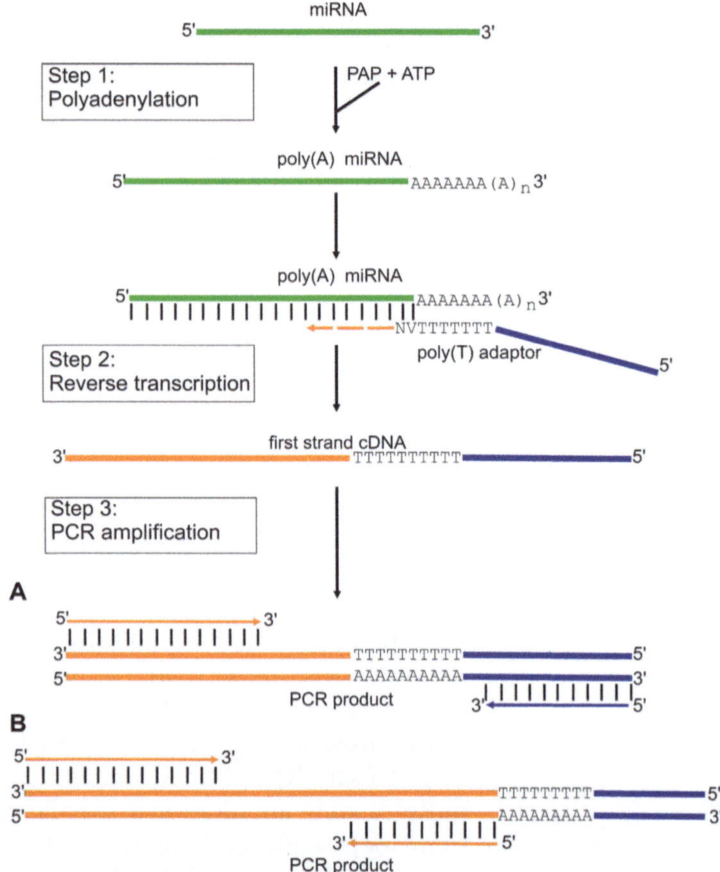

Fig. 2 Schematic description of polyT adaptor-based miRNA qRT-PCR. Step 1: Addition of a polyA-tail to any RNA molecule including miRNAs. Step 2: Reverse transcription with an adaptor primer using an Oligo-dT with unique sequence at the 5′ end for universal reverse primer binding and a NV anchor (N: any base; V: not T) at 3′ end. Step 3: cDNA is used for qRT-PCR quantification of mature miRNA using a specific forward primer and a universal reverse primer (A) and for precursor miRNA using specific forward and reverse primers (B)

2 Materials

Working with RNA is very demanding because of the chemical instability of the RNA and the ubiquitous presence of ribonucleases (RNases). These enzymes are difficult to inactivate, even after prolonged boiling or autoclaving, they can maintain activity. Therefore special precautions should be taken when working with RNA:

- Designate a special area for RNA work only.
- Wear gloves at all times. After putting on gloves, don't touch contaminated surfaces and equipment.

- Whenever possible, use sterile, disposable plastic ware, because these materials are generally RNase-free.
- Glassware should be baked at 180–200 °C for at least 4 h.
- Solutions (water and other solutions) should be treated by adding diethyl pyrocarbonate (DEPC) to 0.1 vol %. Following incubation overnight at room temperature, the treated solutions should be autoclaved for 30 min to remove any trace of DEPC.

PCR assays require special laboratory practices to avoid false-positive amplifications because a single DNA molecule can be detected by this technique. When preparing samples for PCR amplification, the following precautions should be taken:

- Prevent aerosols.
- Wear clean gloves (not previously worn while handling amplified material or used during sample preparation).
- Designate a PCR setup area and never bring amplified PCR product into this area.
- Open and close all sample tubes carefully, try not to splash PCR samples.
- Keep reactions and components capped as much as possible.
- Use ultrapure water (18 MΩ cm at 25 °C).
- Store all solutions in designated areas free of amplified DNA or artificial templates.

2.1 Tissue

1. Collect desired tissues and snap freeze in liquid nitrogen and store at −80 °C.

2.2 RNA Extraction

1. RNAqueous®-Micro, Micro Scale RNA Isolation Kit (Ambion).

2.3 cDNA Synthesis

1. TaqMan® MicroRNA Assays (ABI) for specific miRNAs.
2. TaqMan® MicroRNA Reverse Transcription Kit.
3. miScript Reverse Transcription Kit (Qiagen).

2.4 qRT-PCR

1. TaqMan® MicroRNA Assays (ABI) for specific miRNAs (*see* Subheading 2.3).
2. TaqMan® Universal PCR Master Mix 2×, No AmpErase® UNG (ABI).
3. miScript SYBR Green PCR Kit (Qiagen).
4. miScript Precursor Assays (Qiagen).

2.5 Equipment

1. Real-time thermal cycler compatible with the selected technology.
2. Spectrophotometer for low sample volumes (1 μL).
3. Agilent 2100 Bioanalyzer (Agilent Technologies).

3 Methods

3.1 Sample Collection and Microdissection (See Note 1)

1. Place surgical tissues resection specimen immediately on ice, and subsequently snap freeze and store at −80 °C.
2. Prepare 5 μm frozen sections from tissue blocks, place briefly in RNase-free ethanol, stain with hematoxylin and eosin (H&E) and let sections to be diagnosed by a pathologist.
3. Section serially tissue blocks containing the piece of interest or cell types (10 μm sections). Stain the slides with H&E and immediately store at −20 °C.
4. Microdissect manually tissue section under a microscope (e.g., BH2, Olympus) using a sterile injection needle (size 0.65 × 25 mm) (*see* **Note 2**).
5. Place microdissected cells into a 500 μL microfuge tube containing 100 μL of lysis solution from the RNAqueous®-Micro Kit (*see* **Note 3**). Make sure that the sample is completely immersed in lysis solution.

3.2 RNA Purification

We recommend the RNAqueous-Micro® Kit for RNA isolation.

1. Following incubation of the lysate at 42 °C for 30 min, perform all the following steps according to the manual of the manufacturer (*see* **Note 4**).
2. Omit DNase step if amount of total RNA is limited.

3.3 Evaluation of RNA Quality and Quantity

Quantitative evaluation of the isolated RNA should be performed using a spectrophotometer which can handle small sample volumes (1 μL) such as the NanoDrop spectrophotometer. The assessment of RNA quality can be performed using a BioAnalyzer. This is an automated bio-analytical device using microfluidics technology that provides electrophoretic separations of the sample and analyzes RNA using the same principles as denaturing agarose gels but requires only 1 ng of total RNA. Profiles generated on BioAnalyzer yield RNA integrity and ribosomal ratios. The ratio of the two rRNA 18S and 28S is expected to be 1.8 for good quality RNA. The BioAnalyzer software generates also the RNA Integrity Number (RIN). The RIN is developed to extract information about RNA sample integrity directly from the electrophoretic trace of the sample. The assigned RIN score can vary from 1 to 10 where 10 indicates maximum integrity and quality of the RNA (*see* **Note 5**) (Fig. 3).

1. Determine RNA quantity using a UV spectrometer (such as NanoDrop) (*see* **Note 6**).
2. Use 1 ng per sample for the assessment of RNA integrity using BioAnalyzer LabChip®.

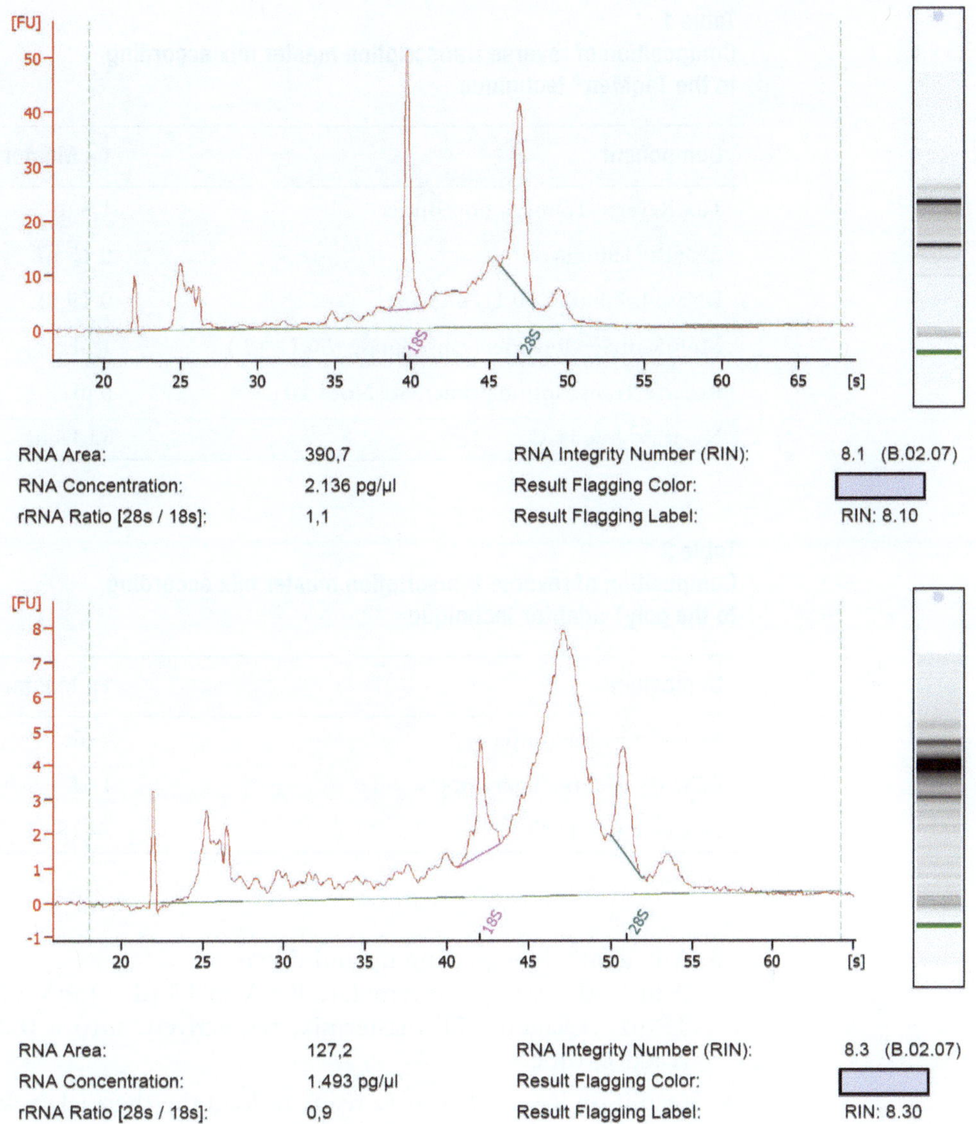

Fig. 3 Total RNA quality analysis via BioAnalyzer. Shown are representative electropherograms and gel-like images from tissue preparations using the RNA 6000 Pico LabChip kit and the 2100 BioAnalyzer. *RIN* RNA integrity value

3.4 cDNA Synthesis for Mature miRNA

1. Adjust RNA to a final concentration of 5 ng/µL and place on ice (*see* **Notes 7** and **8**).
2. Prepare a reverse transcription master mix (RT master mix) for the selected strategy (*see* **Note 9**):

   TaqMan Reverse Transcription Kit components for stem–loop qRT-PCRs (*see* Table 1).

   Qiagen miScript Reverse Transcription Kit for polyT adaptor qRT-PCRs (*see* Table 2).

Table 1
Composition of reverse transcription master mix according to the TaqMan® technique

Component	1× Master mix
10× Reverse Transcription Buffer	1.5 μL
dNTPs (100 mM)	0.15 μL
RNase Inhibitor (20 U/μL)	0.19 μL
MultiScribe™ Reverse Transcriptase (50 U/μL)	1 μL
Reverse Transcription primer (*see* **Note 10**)	3 μL
Nuclease-free H_2O	ad 13 μL

Table 2
Composition of reverse transcription master mix according to the polyT adaptor technique

Component	1× Master mix
5× miScript RT Buffer	4 μL
miScript Reverse Transcriptase Mix	1 μL
Nuclease-free H_2O	ad 18 μL

3. Mix gently by pipetting up and down.
   Add 2 μL (10 ng) of template RNA to 13 μL (TaqMan®) or 18 μL (Qiagen) RT mastermix, respectively, into a 0.2 mL reaction tube.
4. Keep it on ice until you are ready to load the thermal cycler.
   Stem–loop qRT-PCR (TaqMan®): Before loading the thermal cycler, incubate samples 5 min on ice. Use then the following program:

16 °C	30 min
42 °C	30 min
85 °C	5 min

   PolyT adaptor qRT-PCR (Qiagen): Use the following program (preincubation is not required):

37 °C	60 min
95 °C	5 min

5. Store the RT reaction product at −20 °C if you do not immediately continue with the PCR amplification step.

3.5 cDNA Synthesis for Precursor miRNA (See Note 11)

1. Adjust RNA to a final concentration of 50 ng/μL and place on ice (*see* **Notes 7** and **8**).
2. Prepare the RT master mix using the Qiagen miScript Reverse Transcription Kit for polyT adaptor qRT-PCRs (*see* Table 3).
3. Mix gently by pipetting up and down.

   Add 2 μL (100 ng) of template RNA to the RT master mix into a 0.2 mL reaction tube.
4. Keep it on ice until you are ready to load the thermal cycler. Use the following program:

37 °C	60 min
95 °C	5 min

5. Store the RT reaction product at −20 °C if you do not immediately continue with the PCR amplification step.

3.6 qRT-PCR for Mature miRNA (TaqMan®)

Each cDNA sample should be analyzed in triplicate. In addition, a target molecule for normalization should be chosen and processed in parallel (*see* **Note 12**). Non-RT (without reverse transcriptase) and no-template controls (NTCs) are also assayed in triplicate (*see* **Note 13**).

1. Dilute cDNA 1:5 (+60 μL H_2O).
2. Prepare a qRT-PCR reaction mix specific for each primer set according to Table 4 in a 1.5 mL reaction tube (*see* **Note 14**). Add 10–20 % excess volume (*see* **Note 15**).
3. Cap the tube, invert several times to mix and centrifuge briefly.
4. Transfer 10 μL of the qRT-PCR reaction mix into one well on a 96-well PCR plate.
5. Add 5 μL of diluted template cDNA (respectively, H_2O for NTCs) per well.
6. Seal plate and vortex carefully.
7. Briefly centrifuge the plate at 450–500 ×*g*.
8. Load the plate into the instrument and run the qRT-PCR using the following program:

95 °C	10 min	
95 °C	15 s	40x
60 °C	1 min	40x

Table 3
Composition of reverse transcription master mix according to the polyT adaptor technique

Component	1× Master mix
5× miScript RT Buffer	4 μL
miScript Reverse Transcriptase Mix	1 μL
Nuclease-free H_2O	ad 18 μL

Table 4
Composition of qRT-PCR reaction mix according to the TaqMan® technique

Component	1× Master mix (μL)
H_2O	1.75
20× TaqMan® MicroRNA Assay	0.75
2× TaqMan® Universal PCR Master Mix, No AmoErase UNG	7.5

Table 5
Reaction setup for qRT-PCR for detection of mature miRNAs according to the polyT adaptor technique

Component	1× Master mix (μL)
2× QuantiTect SYBR Green PCR Master Mix	10
10× miScript Universal Primer	2
10× miScript Primer Assay	2
H_2O	4

3.7 polyT Adaptor qRT-PCR for Mature miRNA (Qiagen)

1. Prepare a qRT-PCR reaction mix specific for each primer set according to Table 5 in a 1.5 mL reaction tube. Add 10–20 % excess volume (*see* **Note 15**).
2. Cap the tube, invert several times to mix and centrifuge briefly.
3. Transfer 18 μL of the qRT-PCR reaction mix into one well on a 96-well PCR plate.
4. Add 2 μL of template cDNA (respectively, H_2O for NTCs) per well.
5. Seal plate and vortex carefully.
6. Briefly centrifuge the plate at 450–500×*g*.

7. Load the plate into the instrument and run the qRT-PCR using the following program:

95 °C	15 min	
94 °C	15 s	40x
55 °C	30 s	
70 °C	30 s	

It is strongly recommended to perform a melting curve (70–85 °C, read every 0.5 °C for 3 s) analysis at the end of the cycling program (*see* **Note 16**).

3.8 qRT-PCR for Precursor miRNA

1. Prepare a qRT-PCR reaction mix according to Table 6 in a 1.5 mL reaction tube. Add 10–20 % excess volume.
2. Cap the tube, invert several times to mix and centrifuge briefly.
3. Transfer 18 μL of the qRT-PCR reaction mix into one well on a 96-well PCR plate.
4. Add 2 μL of template cDNA (respectively, H_2O for NTCs) per well.
5. Seal plate and vortex carefully.
6. Briefly centrifuge the plate at 450–500 × *g*.
7. Load the plate into the instrument and run the qRT-PCR using the following program:

95 °C	15 min	
94 °C	15 s	40x
55 °C	30 s	
70 °C	30 s	

Table 6
Reaction setup for qRT-PCR for detection of precursor miRNAs according to the polyT adaptor technique

Component	1× Master mix (μL)
H_2O	6
10× miScript Precursor Assay (*see* **Note 17**)	2
2× QuantiTect SYBR Green PCR Master Mix	10

It is strongly recommended to perform a melting curve (70–85 °C, read every 0.5 °C for 3 s) analysis at the end of the cycling program (*see* **Note 18**).

3.9 Relative Quantification of miRNAs

Relative quantification is the most commonly used analysis strategy determining changes in steady-state gene expression levels in a given sample relative to the expression of a reference gene. The data output is expressed as a fold-change or a fold-difference of expression levels. We recommend to calculate the relative expression ratio of a target gene using the method developed by M.W. Pfaffl [5]. PCR efficiency can be established either as described by Pfaffl et al. or using algorithms such as implemented in the LinRegPCR software [6] (for more detailed information refer to the specific pages on http://www.Gene-Quantification.info edited by Michael W. Pfaffl). A more simplistic approach is the so-called "delta-delta Ct method" (Applied Biosystems) which is based on the assumption that real-time amplification efficiencies of both target and reference gene are close to the ideal efficiency of 2 (the amount of PCR amplicon is doubled in each PCR cycle). The known strong impact of differences in primer efficiencies on the calculated expression results should restrict the usage of this method to quick estimations only.

4 Notes

1. For a more detailed protocol please refer to [4].
2. Manual microdissection is used because, in our hands, the procedure is easier and faster than using laser capture microdissection, besides this technology (LCM) presents challenges, particularly with regard to isolating intact RNA for downstream applications.
3. If necessary, samples could be stored in lysis solution at −80 °C.
4. Make sure that the sample solution is clear otherwise disrupt the sample using a micropestle. However, try to avoid this step because this will reduce the yield of RNA quantity.
5. Although the RIN facilitates the assessment of the RNA quality, it should not be taken as an absolute tool for determining the quality of RNA. As shown in Fig. 3 the RIN does not necessarily reflect the exact level of RNA degradation.
6. The amount of isolated RNA should be initially defined by using a spectrophotometer in order to avoid the overloading of the RNA LabChip®. In addition, the sample concentration read from the BioAnalyzer should be regarded as semiquantitative.

7. Due to the low RNA content in some manually microdissected samples and the resulting low amount of isolated total RNA, it is possible that the RNA needs to be concentrated (i.e., via speed vac).
8. Amount of input RNA can be reduced to 1 ng; however, this may reduce the sensitivity of your assay for low abundant miRNA molecules.
9. Correct for losses during pipetting by increasing volumes in your master mix by 10–20 %.
10. RT is possible in multiplex using more than one miRNA-specific stem–loop primer in one RT reaction. We highly recommend controlling the results obtained from multiplex reactions in a representative experiment using singleplex qRT-PCR for each miRNA. Also include an appropriate number of ncRNA molecules for normalization (*see* also **Note 12**).
11. If qRT-PCR may involve quantification of mature and precursor miRNAs in parallel, use the higher amount of RNA input for cDNA synthesis for detection of mature miRNA as well.
12. Normalization is still an unresolved issue in qRT-PCR due to the lack of endogenous molecules which are expressed at a stable level in any cell/tissue type under all possible biological conditions. Therefore, it is difficult to give a general recommendation for a specific probe for normalization in miRNA qRT-PCR assays. However, we recommend testing at least two to three different ncRNA molecules (miRNAs or snoRNAs) in your sample set in order to choose those molecules being most stable under your experimental conditions.
13. We strongly recommend the use of NTC reactions to evaluate background signal.
14. Keep all TaqMan® MicroRNA Assays protected from light, excessive exposure to light may affect the fluorescent probes.
15. Due to the initial incubation step at 95 °C during the PCR reaction, it is not necessary to keep samples on ice during reaction setup.
16. In addition, we recommend checking the specificity of the PCR product using standard gel electrophoresis.
17. The miScript Precursor Assay contains both a forward and a reverse primer. Do not use the miScript Universal Primer.
18. In addition, we recommend checking the specificity of the PCR product using an electrophoretic gel.

References

1. Schmittgen TD, Lee EJ, Jiang J, Sarkar A, Yang L, Elton TS, Chen C (2008) Real-time PCR quantification of precursor and mature microRNA. Methods 44:31–38
2. Shi R, Chiang VL (2005) Facile means for quantifying microRNA expression by real-time PCR. Biotechniques 39:519–525
3. Chen C, Ridzon DA, Broomer AJ, Zhou Z, Lee DH, Nguyen JT, Barbisin M, Xu NL, Mahuvakar VR, Andersen MR et al (2005) Real-time quantification of microRNAs by stem-loop RT-PCR. Nucleic Acids Res 33:e179
4. Luttges J, Hahn SA, Heidenblut AM (2010) Manual microdissection combined with antisense RNA-longSAGE for the analysis of limited cell numbers. Methods Mol Biol 576:135–154
5. Pfaffl MW (2001) A new mathematical model for relative quantification in real-time RT-PCR. Nucleic Acids Res 29:e45
6. Ramakers C, Ruijter JM, Deprez RH, Moorman AF (2003) Assumption-free analysis of quantitative real-time polymerase chain reaction (PCR) data. Neurosci Lett 339:62–66

Chapter 11

Cellular MicroRNA Sensors Based on Luciferase Reporters

Colleen M. Connelly and Alexander Deiters

Abstract

Recently, microRNAs (miRNAs) have been linked to a variety of human diseases including cancer and viral infections. Small molecule modifiers of miRNAs could represent new therapeutic agents and be used as tools for elucidating the biological roles of miRNAs. In order to identify small molecule modifiers of miRNAs, functional assays for specific miRNAs must be developed and optimized. Here, we report the construction of a luciferase reporter assay for miRNA miR-122 function and the development of a stable Huh7 cell line that can be used for high-throughput screening of small molecule miR-122 inhibitors. The steps described here can be applied not only to Huh7 cells and miR-122 but also to virtually any cell line and miRNA combination.

Key words MicroRNA, Cell-based assay, Luciferase, High-throughput screen

1 Introduction

MicroRNAs (miRNAs), single-stranded RNAs of approximately 22 nucleotides, regulate gene expression in a sequence-specific fashion by binding partially complementary sequences in the 3′-untranslated region (3′-UTR) of target messenger RNAs (mRNAs) [1]. MiRNAs are transcribed from the genome into primary miRNAs (pri-miRNAs) and undergo several posttranscriptional processing steps via a dedicated miRNA pathway to produce mature miRNAs. MiRNAs down-regulate gene function by inhibiting translation, accelerating the degradation of the target mRNA, or mediating deadenylation of the mRNA [1, 2]. Almost 2,000 human miRNAs are currently listed in miRBase [3, 4], and it is estimated that they are involved in the control of more than 30 % of all genes [5]. Biological processes regulated by miRNAs include embryonal development, differentiation, apoptosis, and proliferation [6], and the targets of miRNAs range from signalling proteins and transcription factors to RNA binding proteins [7]. Recently, the aberrant expression of certain miRNAs has been linked to a variety of human diseases, including cancer, cardiovascular diseases,

Christoph Arenz (ed.), *miRNA Maturation: Methods and Protocols*, Methods in Molecular Biology, vol. 1095,
DOI 10.1007/978-1-62703-703-7_11, © Springer Science+Business Media New York 2014

immune disorders, and viral infections [8–12]. Small molecule modifiers of miRNA function [13–15] could serve as important tools for elucidating the biological roles of miRNAs and as lead structures for the development of potential new therapeutic agents. In order to identify small molecule modifiers of miRNAs, first functional assays for specific miRNAs must be developed. Here, we report the development of an assay for miRNA miR-122 function based on a luciferase reporter.

MiRNA miR-122 is involved in the regulation of lipid and cholesterol metabolism and is the most abundant miRNA in the liver [16]. Furthermore, miR-122 is down-regulated in hepatocellular carcinoma (HCC) and targets the anti-apoptotic Bcl-2 family member Bcl-w [17]. It was discovered that miR-122 is required by the hepatitis C virus for replication and infectious virus production through interaction with the viral genome [18]. The HCV genome contains two miR-122 target sites in the 5′ noncoding region of the virus and miR-122 recognition results in the up-regulation of viral RNA production [19]. Moreover, miR-122 may stimulate HCV translation [20] through interactions with the HCV internal ribosome entry site (IRES) and argonaute proteins [21]. It was recently reported that knockdown of miR-122 with antisense agents resulted in a decrease in HCV RNA replication in human liver cells [18] and reduced HCV levels in chronically infected primates [22]. These promising results suggest that small molecule inhibitors that target miR-122 could provide a new approach for antiviral therapy. The development of a functional assay followed by a high-throughput screen has the potential to deliver such molecules.

A reporter assay for mature miR-122 function was constructed based on the psiCHECK-2 (Promega) reporter plasmid [15]. The psiCHECK-2 vector has a multi-cloning site at the 3′ terminus of the *Renilla* luciferase gene where a miRNA target sequence can be inserted. Here, the complementary sequence of mature miR-122 was cloned between the *Pme*I and *Sgf*I restriction sites, generating the reporter construct psiCHECK-miR122. After transfection of the psiCHECK-miR122 construct into cells expressing endogenous miR-122, *Renilla* luciferase expression is inhibited (Fig. 1a). In the presence of a miR-122 inhibitor, the *Renilla* luciferase expression will be restored, leading to an increased luciferase signal (Fig. 1b). Utilizing a reporter assay that results in an increased luciferase signal in the presence of an active miR-122 inhibitor rules out false-positive hits due to compound toxicity, which can occur in an assay based on a decrease in reporter signal [23]. The ability of the reporter to detect endogenous miR-122 was confirmed by transiently transfecting the generated psiCHECK-miR122 construct into Huh7 human hepatoma cells [15]. The assay was further validated by co-transfection of a miR-122 antagomir antisense agent as a positive control. The empty psiCHECK-2 vector without the

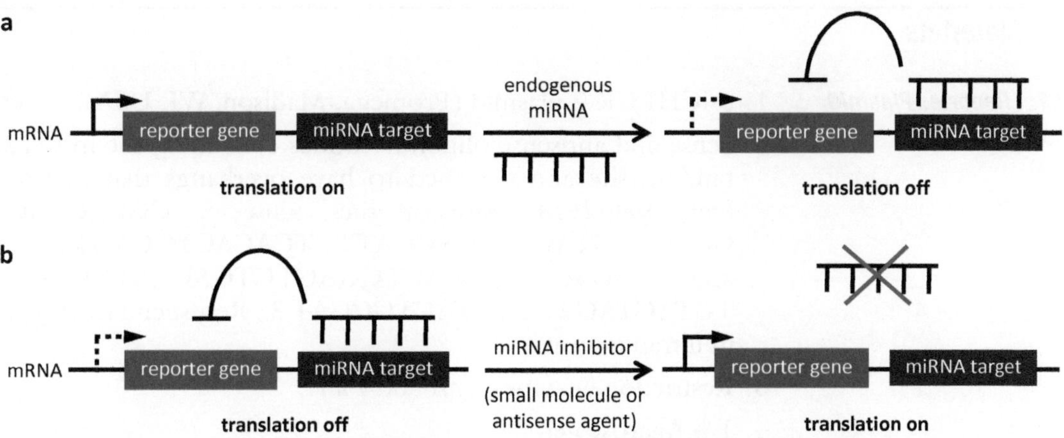

Fig. 1 Design of the microRNA miR-122 reporter assay. (**a**) The developed luciferase reporter can detect the presence of a functional mature miR-122 through repression of the *Renilla* luciferase signal. (**b**) The presence of a small molecule inhibitor of the specific miRNA or an antagomir antisense agent inhibits miRNA function and restores reporter gene (luciferase) expression

miR-122 target sequence (psiCHECK-control) was also transfected into Huh7 cells as a negative control. The relative luciferase signal of the psiCHECK-miR122 construct in Huh7 cells was reduced 15-fold when compared to the transiently transfected psiCHECK-control, validating that the psiCHECK-miR122 reporter is a cellular sensor for endogenous miR-122. From the constructed miR-122 functional assay, a stable Huh7 cell line that constitutively expresses the miR-122 reporter system was developed [24]. The stable Huh7-psiCHECK-miR122 cell line was validated using a miR-122 antagomir antisense agent. The assay was further tested at various DMSO concentrations to confirm that the solvent has no effect on the luciferase assay results. Using the Huh7-psiCHECK-miR122 cell line and a miR-122 antagomir antisense agent, a Z′ factor sufficient for high-throughput screening was obtained [25]. Using a stable cell line instead of a transient transfection will not only be more cost efficient and reduce screening time, but will also remove variation associated with transient transfection efficiency and additional manipulations, making it more suitable for a robust high-throughput small molecule screen.

The steps to create both the functional assay [(1) construction of the reporter plasmid; (2) assessment of the reporter as a miR-122 sensor by transient transfection] and the stable cell line [(1) determination of the G418 concentration for selection; (2) transfection of the stable reporter cell line; (3) selection of the stable reporter cell line; (4) determining the effect of DMSO on the reporter cell line; and (5) determining the Z′ factor] are reported here and can be applied not only to Huh7 cells and miR-122 but also to any other cell line and miRNA combination.

2 Materials

2.1 Reporter Plasmid Construction

1. psiCHECK-2 plasmid (Promega, Madison, WI, USA).
2. Sense and antisense oligonucleotides containing the miR-122 binding site and designed to have overhangs that generate (cut) *Sfg*I/*Pme*I restriction sites: sense- 5′ CGCAGTAGA GCTCTAGTACAAACACCATTGTCACACTCCAGTTT 3′ and antisense- 5′ AAACTGGAGTGTGACAATGGTGTT TGTACTAGAGCTCTACTGCGAT 3′. Resuspend to 100 μM in ultrapure water.
3. Restriction enzymes: *Sgf*I and *Pme*I.
4. 1 % Agarose gel.
5. 1× TBE buffer: 89 mM Tris base, 89 mM boric acid, 2 mM EDTA.
6. QIAquick Gel Extraction Kit (Qiagen, Valencia, CA, USA).
7. T4 DNA ligase (New England Biolabs, Ipswich, MA, USA).
8. NovaBlue competent *Escherichia coli* cells.
9. psiCHECK-2 sequencing primer: 5′ GCTAAGAAGTTCCCT 3′.

2.2 General Cell Culture

1. Dulbecco's Modified Eagle's Medium (DMEM, *see* **Note 1**): Add DMEM/HIGH with L-glutamine powder (11.8 g) and sodium bicarbonate (3.26 g) to ultrapure water (880 mL). Mix and adjust the pH to 7.4 (*see* **Note 2**). Add fetal bovine serum (100 mL) and penicillin/streptomycin (50× solution, 20 mL) and filter sterilize. Store at 4 °C.
2. DMEM supplemented with 500 μg/mL G418: Add DMEM/HIGH with L-glutamine powder (11.8 g), sodium bicarbonate (3.26 g), and G418 disulfate (250 mg) to ultrapure water (880 mL). Mix and adjust the pH to 7.4. Add fetal bovine serum (100 mL) and penicillin/streptomycin (50× solution, 20 mL) and filter sterilize. Store at 4 °C.
3. TrypLE Express (*see* **Note 3**): a trypsin replacement (Invitrogen, Grand Island, NY, USA).

2.3 Assessment of the psiCHECK-miR122 Reporter System

1. White clear-bottom 96-well cell culture plates.
2. X-tremGENE siRNA transfection reagent (Roche).
3. Opti-MEM Reduced Serum Medium: Add Opti-MEM Reduced Serum Medium powder (6.8 g; Invitrogen) and sodium bicarbonate (1.2 g) to ultrapure water (450 mL). Mix and adjust the pH to 7.3. Bring the volume to 500 mL with ultrapure water and filter sterilize. Store at 4 °C.
4. Dual Luciferase Reporter Assay Kit (Promega): Prepare 1× Passive Lysis Buffer, Luciferase Assay Reagent II, and Stop & Glo Reagent according to the manufacturer's protocol.

5. Biotek Synergy 4 microplate reader (Biotek) or an equivalent microplate reader capable of reading luminescence (*see* **Note 4**).
6. miR-122 antagomir antisense agent: 5′ ACAAACACCAUUGUCACACUCCA 3′ with 2′-OMe modified bases and phosphorothioate linkages. Dissolve to 100 μM in ultrapure water.
7. 1× Phosphate-buffered saline (PBS): 137 mM NaCl, 2.7 mM KCl, 10 mM Na_2HPO_4, 2 mM KH_2PO_4, pH 7.4, sterilize.

2.4 Determination of the G418 Concentration for Selection of Stable Cells

1. 12-well cell culture plates.
2. G418 disulfate: Add G418 disulfate (200 mg) to ultrapure water (1 mL). Filter sterilize.
3. Cell Titer-Glo Luminescent Cell Viability Assay (Promega): Prepare the Cell Titer-Glo Reagent according to the manufacturer's protocol.

2.5 Transfection of the Huh7-psiCHECK-miR122 Reporter Cell Line

1. pcDNA3 plasmid (Invitrogen).
2. Restriction enzymes: *Bam*HI and *Bgl*II.
3. 15 cm, 6-well, 48-well cell culture plates.
4. Lipofectamine 2000 transfection reagent (Invitrogen).
5. Cloning cylinders (Corning, Tewksbury, MA, USA).

2.6 Effect of DMSO on the Huh7-psiCHECK-miR122 Reporter Cell Line

1. Cell culture grade DMSO: Filter sterilize.

3 Methods

3.1 Reporter Plasmid Construction

1. Sequentially digest the psiCHECK-2 plasmid (1 μg) with *Sgf*I (10 units, 50 μL reaction) followed by *Pme*I (10 units; 50 μL reaction) at 37 °C for 2 h and heat inactivate at 75 °C for 20 min.
2. Separate the digested backbone on a 1 % agarose gel in TBE buffer at 90 V for 30 min. Excise the digested backbone and purify using a QIAquick Gel Extraction Kit (see manufacturer's protocol).
3. Hybridize the insert DNA containing the miR-122 binding site by combining 50 μL of both the sense (5′CGCAGTAGAGCTCTAGTACAAACACCATT GTCACACTCCAGTTT 3′) and the antisense (5′AAACTG GAGTGTGACAATGGTGTTTGTACTAGAGCTC TACTGCGAT 3′) oligonucleotides (1 μM) and incubate at 90 °C for 5 min followed by cooling to 4 °C over 5 min and finally, incubate at 4 °C for 60 min.

4. Ligate the annealed insert with T4 ligase (200 units, 10 μL reaction, 1:10 vector/insert ratio) into the digested psiCHECK-2 vector at 4 °C overnight.
5. Transform the ligation reaction into NovaBlue competent cells and plate on an LB agar plate containing ampicillin (50 μg/mL). Incubate at 37 °C overnight (*see* **Note 5**).
6. The construction of the psiCHECK-miR122 vector is confirmed by sequencing (sequencing primer: 5′ GCTAAGAA GTTCCCT 3′).

3.2 General Cell Culture

All cell culture experiments using the Huh7 human hepatoma cell line should be performed in standard Dulbecco's Modified Eagle's Medium (DMEM) supplemented with 10 % fetal bovine serum (FBS) and 1 % penicillin/streptomycin and maintained at 37 °C in a 5 % CO_2 atmosphere. The Huh7-psiCHECK-miR122 cell line should be cultured in DMEM supplemented with 10 % FBS, 500 μg/mL of G418 (*see* **Note 6**), and 1 % penicillin/streptomycin and maintained at 37 °C in a 5 % CO_2 atmosphere. All Huh7 cell lines can be passaged using TrypLE Express (trypsin replacement).

3.3 Assessment of the psiCHECK-miR122 Reporter System

To validate the psiCHECK-miR122 reporter as a sensor for miR-122, the plasmid is transfected into Huh7 cells, which express high levels of miR-122, thus leading to a decrease in *Renilla* luciferase expression compared to the psiCHECK-2 control plasmid. In order to verify that the *Renilla* luciferase expression is restored by inhibition of mature miR-122, Huh7 cells are co-transfected with the psiCHECK-miR122 reporter and a miR-122 antagomir antisense agent.

1. Passage Huh7 cells into a white clear-bottom 96-well plate and grow overnight at 37 °C until they reach 60 % confluency.
2. Transfect either the psiCHECK-control plasmid (the original psiCHECK-2 plasmid containing no known miRNA binding site) or the psiCHECK-miR122 plasmid (0.5 μg) with or without the miR-122 antagomir antisense agent (50 pmol) using X-tremGENE siRNA transfection reagent (2:1 reagent/DNA ratio) in Opti-Mem media (200 μL total volume) according to the manufacturer's protocol. Perform all transfections in triplicate for statistical analysis.
3. Incubate the cells at 37 °C for 4 h.
4. Remove the Opti-Mem media and replace with standard DMEM growth media (200 μL).
5. Incubate the cells at 37 °C for 48 h.
6. Remove the media and rinse the cells with 1× PBS (50 μL).
7. Lyse the cells by adding 1× Passive Lysis Buffer (25 μL; supplied in Dual Luciferase Reporter Assay Kit) and shake for 15 min at room temperature.

8. Assay the lysed cells using a Dual Luciferase Reporter Assay Kit according to the manufacturer's protocol and record the luminescence on a microplate reader. In short, for each well, dispense the Luciferase Assay Reagent II (100 μL), read firefly luciferase activity with a measurement time of 10 s and a delay time of 2 s, dispense the Stop & Glo Reagent (100 μL), read *Renilla* luciferase activity with a measurement time of 10 s and a delay time of 2 s.
9. Calculate the ratio of *Renilla* to firefly luciferase expression for each well to provide relative luciferase units (RLUs). The average and standard deviation for each of the triplicates is calculated from the RLU values.

3.4 Determination of the G418 Concentration for Selection of Stable Cells

Transient transfection of the psiCHECK-miR122 reporter has several limitations for high-throughput screening such as the high cost of transfection reagents, extensive transfection procedures, and increased variations between different plates and different days. To create a robust high-throughput assay for small molecule inhibitors of miR-122, a stable cell line that constitutively expresses a miR-122 reporter system is necessary to reduce the number of manipulations, to increase the density of wells screened per plate, and to enhance the reproducibility of the assay. The following steps to create the stable cell line can be applied not only to Huh7 cells but also to any other cell line by adjusting the G418 selective pressure.

1. Passage Huh7 cells into a 12-well plate and grow overnight at 37 °C until they reach 30 % confluency.
2. Remove the growth media and treat the cells with DMEM (1 mL) supplemented with G418 (0, 50, 100, 150, 200, 300, 400, 500, 600, 700, 800, or 1,000 μg/mL).
3. Incubate the cells at 37 °C for 48 h. Visually inspect the cells and note the concentration that causes ~80 % cell death after 48 h.
4. Continue treatment by replacing the media every 2–3 days with freshly prepared selective media containing the same concentration of G418. Incubate the cells at 37 °C for 14 days after the first G418 treatment.
5. After 14 days, assay the cells for viability using a Cell Titer-Glo Luminescent Cell Viability Assay according to the manufacturer's protocol and record the luminescence on a microplate reader. In short, equilibrate the plate to room temperature for approximately 30 min. Add a volume of Cell Titer-Glo Reagent equal to the growth media volume. Mix for 2 min and allow the plate to incubate at room temperature for 10 min. Record the luminescence with a measurement time of 1 s. For Huh7 cells, selection with 500 μg/mL G418 is sufficient for 80 % cell death after 48 h and 100 % cell death after 14 days (*see* **Note 7**).

3.5 Transfection of the Huh7-psiCHECK-miR122 Reporter Cell Line

The psiCHECK-miR122 reporter does not contain a selectable marker for the generation of a stably transfected mammalian cell line. Stable transfectants can be generated by co-transfection of psiCHECK-miR122 with the pcDNA3 plasmid which encodes a neomycin resistance marker. Linearization of both plasmids in a noncoding area prior to transfection ensures the integrity of the essential genes within the plasmid including the luciferase reporters and neomycin resistance, which could otherwise be destroyed by a random digestion within the cell [26].

1. Linearize the psiCHECK-miR122 and pcDNA3 plasmids with *Bam*HI (20 units, 100 μL reaction) and *Bgl*II (20 units, 100 μL reaction), respectively, and purify the DNA by ethanol precipitation. Confirm that the plasmids have been linearized by analyzing the DNA on a 1 % agarose gel (TBE buffer, 90 V for 30 min).
2. Passage Huh7 cells into a 6-well plate and grow overnight at 37 °C until they reach 90 % confluency.
3. Cotransfect one well of Huh7 cells with the linearized psiCHECK-miR122 (2.0 μg) and linearized pcDNA3 (0.25 μg) using Lipofectamine 2000 transfection reagent (3:1 volume Lipofectamine 2000/amount psiCHECK-miR122 DNA ratio) in Opti-Mem media (2 mL total volume) according to the manufacturer's protocol. The transfection conditions should be optimized by transient transfection prior to generating the stable cell line.
4. Incubate the cells at 37 °C for 5 h. Remove the media and replace with standard DMEM growth media.
5. Incubate the cells for 48 h. Remove the media, detach the cells using TrypLE Express (500 μL), and passage the cells into a 15 cm cell culture plate.
6. Start the selection process by treating the cells with DMEM growth media containing 500 μg/mL of G418. Incubate the cells at 37 °C for approximately 14 days or until resistant colonies begin to grow. The media should be removed and replaced with fresh selective media every 2–3 days.
7. After 14 days, remove the media and isolate the resistant clones using cloning cylinders. Add TrypLE Express (50 μL) to each cloning ring to detach the cells. Transfer the suspended cells to one well of a 48-well plate containing DMEM growth media (500 μL) supplemented with 500 μg/mL of G418.
8. Incubate the selected clones at 37 °C until the wells are confluent enough to passage.
9. Passage each clone into one well of a 6-well plate in order to grow enough cells to assay for luciferase expression.

10. From the 6-well plate, passage each clone in triplicate into a white clear-bottom 96-well plate at 10,000 cells/well and grow overnight at 37 °C.
11. Remove the media and rinse the cells with 1× PBS (50 μL).
12. Lyse the cells by adding 1× Passive Lysis Buffer (25 μL) and shake for 15 min at room temperature.
13. Assay the cells for luciferase expression using a Dual Luciferase Reporter Assay Kit as described previously (*see* Subheading 3.3) and record the luminescence on a microplate reader (*see* **Note 8**).

3.6 Selection of a Stable Huh7-psiCHECK-miR122 Reporter Cell Line

Promising clones obtained from the G418 selections, which have high firefly luciferase readings and low relative luciferase signals, are further validated by transfection with a miR-122 antagomir antisense agent to verify that the *Renilla* luciferase expression responds to a loss of miR-122 function.

1. Passage each clone into a white clear-bottom 96-well plate at 10,000 cells/well and incubate overnight at 37 °C. Transfect each clone in triplicate with and without the miR-122 antagomir antisense agent (250 pmol) using X-tremeGENE siRNA transfection reagent (2 μL/well) in Opti-Mem media.
2. Incubate the cells at 37 °C for 4 h. Replace the transfection media with standard DMEM growth media (without G418).
3. After a 48 h incubation, remove the media, wash the cells with 1× PBS, lyse the cells, and assay with a Dual Luciferase Reporter Assay Kit as described previously (*see* Subheading 3.3). The stable cell line should be chosen based on the level of luciferase expression and the response to miR-122 antagomir antisense agent transfection. For the miR-122 antagomir transfection, a minimum signal to background of 3 is required for a robust assay [27].

3.7 Effect of DMSO on the Huh7-psiCHECK-miR122 Reporter Cell Line

To validate that the Huh7-psiCHECK-miR122 reporter cell line can be used in a high-throughput assay for small molecule inhibitors of miR-122, the cells need to be tested with increasing concentrations of DMSO since small molecule libraries are commonly stored in DMSO as a solvent [28, 29]. In the Dual Luciferase Assay, the DMSO should have minimal effect on luciferase expression and cell viability at the concentration to be used for small molecule screening, typically below 1 % for cellular assays [28].

1. Passage the selected Huh7-psiCHECK-miR122 cells at 10,000 cells per well into a white clear-bottom 96-well plate and incubate at 37 °C overnight.
2. Treat the cells with DMSO (0, 0.1, 0.5, 1, and 2 %) in standard DMEM growth media and incubate the cells at 37 °C for 48 h.
3. Remove the media, wash the cells with 1× PBS, lyse the cells, and assay with a Dual Luciferase Reporter Assay Kit as described previously (*see* Subheading 3.3).

3.8 Determination of the Z′ factor

Any developed assay needs to be further validated for HTS-readiness by calculating the statistical parameter Z′, which represents a quantitative and well-established measure of the quality of the assay [25]. The Z′ factor takes into account the precision of measuring the maximum and minimum control signals in replicate wells. The assay should be repeated over multiple days to ensure reproducibility. A Z′ factor of 0.5 or above represents an excellent assay and is sufficient for high-throughput screening [25, 30, 31].

1. Passage the Huh7-psiCHECK-miR122 cells at 10,000 cells per well in white clear-bottom 96-well plates. After an overnight incubation, transfect the cells with the miR-122 antagomir antisense agent (0 or 250 pmol, negative and positive control, respectively) using the X-tremeGENE siRNA transfection reagent (2 μL/well) in Opti-Mem media as previously described.
2. Incubate the cells at 37 °C for 4 h. Replace the transfection media with standard DMEM growth media.
3. Incubate the cells at 37 °C for 48 h, remove the media, wash the cells with 1× PBS, lyse the cells, and assay with a Dual Luciferase Reporter Assay Kit as previously described (*see* Subheading 3.3). Repeat the assay on multiple days to ensure reproducibility.
4. Determine the Z′ factor using the equation $Z' = 1 - (3 \times SD_{\text{positive control}} + 3 \times SD_{\text{negative control}})/(Avg_{\text{positive control}} - Avg_{\text{negative control}})$ [25].

4 Notes

1. The Dulbecco's modified Eagle's medium (DMEM) described here is made from DMEM/HIGH with L-glutamine powder (Hyclone). This growth media can also be purchased as pre-sterilized liquid media to simplify the media preparation. In this case, the pH does not need to be adjusted and only the FBS, penicillin/streptomycin, and G418 disulfate (for selective media) need to be added.
2. Adjust the pH of the cell culture media before sterilization using HCl (6 M) or NaOH (2 M) made with ultrapure water.
3. TrypLE Express is a trypsin replacement that is gentle on cells, maintains healthy cell growth, and achieves faster dissociation. Trypsin EDTA, which is commonly used in cell culture laboratories, can also be used for passaging the Huh7 cell lines.
4. For the Dual Luciferase Reporter Assay, a microplate reader with dispensers is preferable for delivering the luciferase substrates; however, the procedure can be adjusted for microplate readers without dispensers (see manufacturer's protocol).

5. For the cloning of psiCHECK-miR122, a control ligation reaction without insert should also be performed and transformed into NovaBlue competent cells. If there are background colonies on the control plate, an additional digest with *Xho*I can be performed using the ligation products to remove any re-circularized psiCHECK-2 plasmid without insert.
6. The Huh7-psiCHECK-miR122 cell line should be cultured in DMEM supplemented with 10 % FBS, 2 % penicillin/streptomycin, and the concentration of G418 determined by the viability experiment (*see* Subheading 3.4).
7. The 500 μg/mL of G418 stated here can act as a guideline for Huh7 cells; however, the viability experiment should be performed using the cell line of interest (including Huh7 cells) and the specific G418 batch to be used for selection.
8. Because the stable cell line is developed by co-transfection of psiCHECK-miR122 and the pcDNA3 plasmid, some clones will incorporate only the pcDNA plasmid and will therefore show G418 resistance but no luciferase expression.

References

1. Carthew R (2006) Gene regulation by microRNAs. Curr Opin Genet Dev 16:203–208
2. Djuranovic S, Nahvi A, Green R (2012) miRNA-mediated gene silencing by translational repression followed by mRNA deadenylation and decay. Science 336:237–240
3. Kozomara A, Griffiths-Jones S (2011) miRBase: integrating microRNA annotation and deep-sequencing data. Nucleic Acids Res 39:D152–D157
4. Griffiths-Jones S, Grocock RJ, van Dongen S, Bateman A, Enright AJ (2006) miRBase: microRNA sequences, targets and gene nomenclature. Nucleic Acids Res 34:D140–D144
5. Lewis BP, Burge CB, Bartel DP (2005) Conserved seed pairing, often flanked by adenosines, indicates that thousands of human genes are microRNA targets. Cell 120:15–20
6. Appasani K (2008) MicroRNAs: from basic science to disease biology. Cambridge University Press, Cambridge
7. Janga SC, Vallabhaneni S (2011) MicroRNAs as post-transcriptional machines and their interplay with cellular networks. Adv Exp Med Biol 722:59–74
8. Tong AW, Nemunaitis J (2008) Modulation of miRNA activity in human cancer: a new paradigm for cancer gene therapy? Cancer Gene Ther 15:341–355
9. Sevignani C, Calin G, Siracusa L, Croce C (2006) Mammalian microRNAs: a small world for fine-tuning gene expression. Mamm Genome 17:189–202
10. Port JD, Sucharov C (2010) Role of microRNAs in cardiovascular disease: therapeutic challenges and potentials. J Cardiovasc Pharmacol 56:444–453
11. Lindsay MA (2008) microRNAs and the immune response. Trends Immunol 29: 343–351
12. Cullen BR (2011) Viruses and microRNAs: RISCy interactions with serious consequences. Genes Dev 25:1881–1894
13. Deiters A (2010) Small molecule modifiers of the microRNA and RNA interference pathway. AAPS J 12:51–60
14. Gumireddy K, Young D, Xiong X, Hogenesch J, Huang Q, Deiters A (2008) Small-molecule inhibitors of microrna miR-21 function. Angew Chem Int Ed Engl 47:7482–7484
15. Young D, Connelly C, Grohmann C, Deiters A (2010) Small molecule modifiers of microRNA miR-122 function for the treatment of hepatitis C virus infection and hepatocellular carcinoma. J Am Chem Soc 132:7976–7981
16. Esau C, Davis S, Murray S, Yu X, Pandey S, Pear M et al (2006) miR-122 regulation of lipid metabolism revealed by in vivo antisense targeting. Cell Metab 3:87–98
17. Lin C, Gong H, Tseng H, Wang W, Wu J (2008) miR-122 targets an anti-apoptotic gene, Bcl-w, in human hepatocellular carcinoma cell lines. Biochem Biophys Res Commun 375:315–320

18. Jopling C, Yi M, Lancaster A, Lemon S, Sarnow P (2005) Modulation of hepatitis C virus RNA abundance by a liver-specific MicroRNA. Science 309:1577–1581
19. Jopling CL, Schütz S, Sarnow P (2008) Position-dependent function for a tandem microRNA miR-122-binding site located in the hepatitis C virus RNA genome. Cell Host Microbe 4:77–85
20. Henke JI, Goergen D, Zheng J, Song Y, Schüttler CG, Fehr C et al (2008) microRNA-122 stimulates translation of hepatitis C virus RNA. EMBO J 27:3300–3310
21. Roberts AP, Lewis AP, Jopling CL (2011) miR-122 activates hepatitis C virus translation by a specialized mechanism requiring particular RNA components. Nucleic Acids Res 39:7716–7729
22. Lanford R, Hildebrandt-Eriksen E, Petri A, Persson R, Lindow M, Munk M et al (2010) Therapeutic silencing of microRNA-122 in primates with chronic hepatitis C virus infection. Science 327:198–201
23. Davis RE, Zhang YQ, Southall N, Staudt LM, Austin CP, Inglese J et al (2007) A cell-based assay for IkappaBalpha stabilization using a two-color dual luciferase-based sensor. Assay Drug Dev Technol 5:85–103
24. Connelly CM, Thomas M, Deiters A (2012) High-throughput luciferase reporter assay for small-molecule inhibitors of microRNA function. J Biomol Screen 17(6):822–828
25. Zhang J, Chung T, Oldenburg K (1999) A simple statistical parameter for use in evaluation and validation of high throughput screening assays. J Biomol Screen 4:67–73
26. Stuchbury G, Münch G (2010) Optimizing the generation of stable neuronal cell lines via pre-transfection restriction enzyme digestion of plasmid DNA. Cytotechnology 62:189–194
27. Janzen WP (2002) High throughput screening: methods and protocols. Humana, Totowa, NJ
28. Gad SC (2005) Drug discovery handbook. Wiley-Interscience/Wiley, Hoboken, NJ
29. Cheng X, Hochlowski J, Tang H, Hepp D, Beckner C, Kantor S et al (2003) Studies on repository compound stability in DMSO under various conditions. J Biomol Screen 8: 292–304
30. Iversen PW, Eastwood BJ, Sittampalam GS, Cox KL (2006) A comparison of assay performance measures in screening assays: signal window, Z′ factor, and assay variability ratio. J Biomol Screen 11:247–252
31. Inglese J, Johnson RL, Simeonov A, Xia M, Zheng W, Austin CP et al (2007) High-throughput screening assays for the identification of chemical probes. Nat Chem Biol 3:466–479

Chapter 12

Identification of Inhibitors of MicroRNA Function from Small Molecule Screens

Colleen M. Connelly and Alexander Deiters

Abstract

Aberrant expression of microRNAs (miRNAs) has been linked to many human diseases including cancer, immune disorders, heart disease, and viral infections. Thus, small molecule inhibitors of miRNAs have potential as new therapeutic agents, as probes for the elucidation of detailed mechanisms of miRNA function, and as tools for the discovery of new targets for the treatment of human diseases. In order to identify small molecule inhibitors of specific miRNAs, functional assays have been developed and applied to the screening of small molecule libraries. Here, we report the application of a luciferase-based reporter assay of miRNA miR-122 function to the discovery of small molecule miR-122 inhibitors.

Key words MicroRNA, Cell-based assay, Luciferase, High-throughput screen, Small molecule inhibitor

1 Introduction

The aberrant expression of microRNAs (miRNAs) has been linked to many human diseases including cancer, immune disorders, heart disease, and viral infections [1–5]. Given that miRNAs can play critical roles in cancer and other diseases, it is important to understand their biogenesis, regulation, and function. Small molecule inhibitors of miRNAs could serve as the necessary tools for elucidating the regulation and function of specific miRNAs and could be lead structures for the development of new therapeutic agents [6]. Several regulatory tools to control the function of individual miRNAs have previously been developed and applied to functional studies, including antisense oligonucleotides or antagomirs [7, 8], miRNA sponges, and miRNA decoys [9]. However, these nucleic acid-based molecules face difficulties as tools to study miRNA biogenesis and as potential therapeutics. A few of the major drawbacks include the high cost for in vivo studies, susceptibility to enzymatic degradation, poor cellular uptake and systemic distribution, low oral bioavailability, and the potential for immune responses [10, 11].

Christoph Arenz (ed.), *miRNA Maturation: Methods and Protocols*, Methods in Molecular Biology, vol. 1095,
DOI 10.1007/978-1-62703-703-7_12,

Therefore, small molecules may represent a suitable alternative to miRNA regulation, as they are less expensive to produce, readily diffuse across cell membranes, are easily delivered into cells, animals, and humans with high temporal resolution, and are more stable intracellularly. Lastly, small molecules have more ideal drug properties including good solubility, good bioavailability, and metabolic stability [10].

Small molecules have the potential to control miRNA expression on multiple levels of the miRNA pathway such as pre-transcription, transcription, and posttranscription, in contrast to nucleic acid-based reagents that can only regulate miRNA function through a direct interaction with the mature miRNA. Although all three levels of regulation are important, both pre- and posttranscriptional regulation appear to be generally less miRNA-specific, whereas the transcriptional level offers a higher degree of miRNA specificity since transcription factors are presumably involved in the development- and cell-specific regulation of distinct miRNAs [12]. Consequently, small molecules are more versatile probes to study miRNA biogenesis than nucleic acid-based tools.

In recent years, several small molecule screens have been conducted in order to identify compounds that either activate or deactivate miRNAs [13–19]. Most small molecules that have been identified regulate miRNA expression in a general fashion by either inhibiting or activating miRNA processing. Our approach to small molecule regulation of miRNAs focuses on the discovery and development of small molecules that inhibit specific miRNAs in order to use the identified small molecules as molecular probes for those particular miRNAs of interest. We have developed assay systems to discover miRNA-specific inhibitors and have identified several small molecules that could be useful tools for the elucidation of detailed mechanisms of miRNA action and may serve as lead structures for the development of new therapeutic agents. Specifically, we have focused on the miRNAs miR-21 and miR-122.

MiRNA miR-122 is required by the hepatitis C virus (HCV) for replication and infectious virus production [20]. It was recently reported that knockdown of miR-122 with antisense agents resulted in a decrease in HCV RNA replication in human liver cells [20] and reduced HCV levels in chronically infected primates [21], suggesting that small molecule inhibitors that target miR-122 could be a viable avenue for a fundamentally new antiviral therapy.

A reporter assay for mature miR-122 function was constructed based on the psiCHECK-2 (Promega) reporter plasmid [14]. The complementary sequence for mature miR-122 was inserted downstream of the *Renilla* luciferase gene to create the reporter construct psiCHECK-miR122, and a stable Huh7 reporter cell

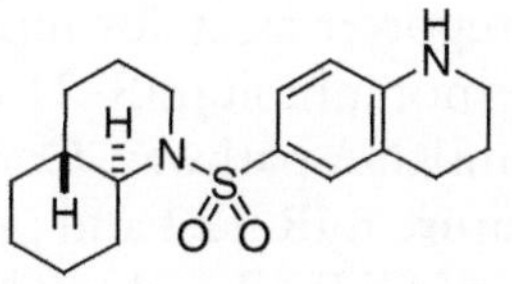

Fig. 1 Small molecule inhibitor of miR-122 discovered in a pilot screen of the NCI Diversity Set II

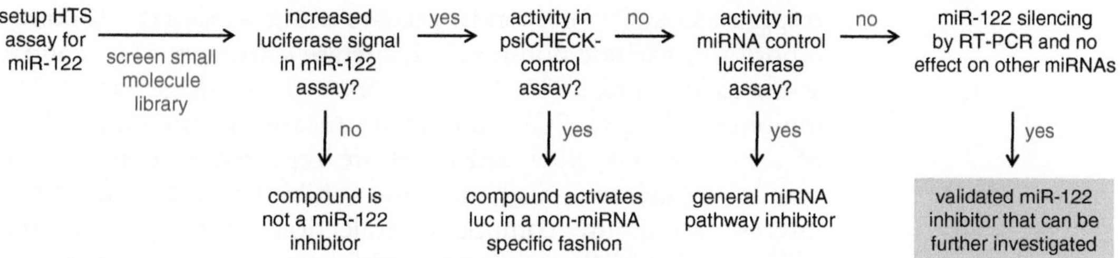

Fig. 2 Steps for the identification of specific miR-122 inhibitors using a high-throughput screen and a set of counter screens

line was generated (*see* Chapter 11 for a detailed discussion of the assay development). In this cell line, the mature miR-122 binds its target sequence and causes a decrease in the *Renilla* luciferase expression. In the presence of a small molecule miR-122 inhibitor, luciferase expression is restored, enabling the identification of small molecule inhibitors of miR-122 function. The developed reporter assay was applied in a pilot screen of small molecule inhibitors of miR-122, followed by several counter screens to: (1) eliminate molecules that increase or decrease luciferase expression in a non-miRNA controlled way (*see* Subheading 3.4), (2) eliminate compounds that act on the miRNA pathway in a general fashion and are not specific for miR-122 (*see* Subheading 3.5), and (3) validate the effect of the small molecules on mature miR-122 levels (*see* Subheading 3.6). The Diversity Set II from the NCI Developmental Therapeutics Program was screened at a 10 μM concentration in a 96-well format and several initial hit compounds were identified as potential miR-122 inhibitors that induced a ≥fivefold increase in the relative luciferase signal.

After further validation following the decision tree outlined in Fig. 2, several of the initial hits were dismissed due to toxicity or non-miRNA-specific interactions with the luciferase reporter. Importantly, the small molecule **1** (Fig. 1) was verified as a potent, low-micromolar miR-122 inhibitor. When Huh7 cells transfected with the psiCHECK-control plasmid were treated with **1**, no change in luciferase signal was observed, indicating that the inhibitor does not increase the luciferase expression in a non-miRNA-specific fashion. Further, the compound showed no effect in a luciferase reporter assay for miR-21 function [13], demonstrating that **1** does not inhibit miR-21 function and is not a general inhibitor of the miRNA pathway. Compound **1** also induced a reduction in both mature miR-122 and primary miR-122 levels, as shown by quantitative RT-PCR [14]. Thus, this pilot screen validated the ability to discover small molecule inhibitors of miR-122 function.

The steps to identify inhibitors of miR-122 by small molecule screening [(1) primary small molecule screen in the psiCHECK-miR122 assay; (2) confirmation of initial hit compounds by dose response assays; (3) secondary assays using a psiCHECK-control reporter to exclude nonspecific luciferase activators; (4) secondary assays using a miR-21 luciferase reporter to exclude general miRNA inhibitors; (5) qRT-PCR experiments measuring intracellular levels of miR-122 and other miRNAs] are reported here and can be applied to identify small molecule inhibitors of any miRNA of interest using any compound collection after an appropriate reporter assay has been constructed.

2 Materials

2.1 General Cell Culture

1. Dulbecco's Modified Eagle's Medium (DMEM, *see* **Note 1**): Add DMEM/HIGH with L-glutamine powder (11.8 g) and sodium bicarbonate (3.26 g) to ultrapure water (880 mL). Mix and adjust the pH to 7.4 (*see* **Note 2**). Add fetal bovine serum (FBS, 100 mL) and penicillin/streptomycin (50× solution, 20 mL) and filter sterilize. Store at 4 °C.
2. DMEM supplemented with 500 μg/mL G418: Add DMEM/HIGH with L-glutamine powder (11.8 g), sodium bicarbonate (3.26 g), and G418 disulfate (250 mg) to ultrapure water (880 mL). Mix and adjust the pH to 7.4. Add FBS (100 mL) and penicillin/streptomycin (50× solution, 20 mL) and filter sterilize. Store at 4 °C.
3. TrypLE Express (Invitrogen): a trypsin replacement (*see* **Note 3**).
4. PCR Mycoplasma Detection Set (Takara Bio, Inc., Shiga, Japan).
5. Huh7 cell line.
6. Huh7-psiCHECK-miR122 stable cell line (Huh7 cells stably transfected with the psiCHECK-miR122 reporter: *see* Chapter 11 for a discussion of this cell line).

2.2 Primary Small Molecule Screen

1. White clear-bottom 96-well tissue culture-treated microtiter plates.
2. psiCHECK-miR122 reporter plasmid (*see* Chapter 11).
3. X-tremGENE siRNA transfection reagent (Roche).

4. miR-122 antagomir antisense agent: 5′ ACAAACACCAUU GUCACACUCCA 3′ with 2′-OMe modified bases and phosphorothioate linkages. Dissolve to 100 μM in ultrapure water.
5. Opti-MEM Reduced Serum Medium (Invitrogen).
6. Compound library in a 96-well format at a working stock dilution of 1 mM in 100 % DMSO. For example, the Diversity Set II small molecule library from the NCI Developmental Therapeutics Program.
7. Cell culture grade DMSO: Filter sterilize.
8. Dual Luciferase Reporter Assay Kit (Promega): Prepare 1× Passive Lysis Buffer, Luciferase Assay Reagent II, and Stop & Glo Reagent according to the manufacturer's protocol.
9. Biotek Synergy 4 microplate reader (Biotek) or an equivalent microplate reader capable of reading luminescence (*see* **Note 4**).
10. 1× Phosphate-buffered saline (PBS): 137 mM NaCl, 2.7 mM KCl, 10 mM Na_2HPO_4, 2 mM KH_2PO_4, pH 7.4, sterilized.

2.3 Secondary psiCHECK-Control Assay

1. psiCHECK-control plasmid: the psiCHECK-2 plasmid containing no known miRNA binding site (Promega, Madison, WI, USA).

2.4 Secondary miR-21 Luciferase Assay

1. psiCHECK-miR21 reporter plasmid (the psiCHECK-2 plasmid containing a miR-21 binding site in the 3′ UTR of the *Renilla* luciferase gene).

2.5 Quantitative RT-PCR

1. 6-well tissue culture-treated plates.
2. mirPremier microRNA Isolation Kit (Sigma Aldrich, St. Louis, MO, USA).
3. Nanodrop ND-1000 spectrophotometer.
4. TaqMan microRNA Reverse Transcription Kit (Applied Biosystems, Carlsbad, CA, USA).
5. TaqMan miRNA assays for miR-122 and miR-21(Applied Biosystems).
6. TaqMan Universal PCR Master Mix, No AmpErase UNG (Applied Biosystems).
7. Bio-Rad MyiQ Real-Time PCR Detection System (Bio-Rad, Hercules, CA, USA).

3 Methods

3.1 General Cell Culture

All cell culture experiments using Huh7 cells should be performed in standard DMEM supplemented with 10 % FBS and 1 % penicillin/streptomycin and maintained at 37 °C in a 5 %

CO_2 atmosphere. The Huh7-psiCHECK-miR122 cell line should be cultured in DMEM supplemented with 10 % FBS, 500 μg/mL of G418, and 1 % penicillin/streptomycin and maintained at 37 °C in a 5 % CO_2 atmosphere. All cell lines can be passaged using TrypLE Express (trypsin replacement). Prior to small molecule screening, the cell lines should be tested for mycoplasma contamination using a Takara PCR Mycoplasma Detection Set according to the manufacturer's protocol. To treat any cell line for possible mycoplasma contamination, cells can be cultured for 2 consecutive weeks with plasmocin (25 mg/mL; Invivogen, San Diego, CA, USA) per the manufacturer's instructions.

3.2 Primary Small Molecule Screen

Once a miR-122 functional assay is developed and optimized, it can be used to screen for small molecule inhibitors of miR-122.

1. Passage Huh7-psiCHECK-miR122 cells into a white clear-bottom 96-well plate at 10,000 cells/well and grow overnight at 37 °C.
2. Transfect the Huh7-psiCHECK-miR122 cells with the miR-122 antagomir antisense agent (50 pmol) in triplicate as a positive control using X-tremGENE siRNA transfection reagent (1 μL) in Opti-Mem media (200 μL total volume) according to the manufacturer's protocol.
3. Incubate the cells at 37 °C for 4 h.
4. Replace the Opti-MEM media with standard DMEM growth media (100 μL).
5. For compound treatment, replace the growth media with standard DMEM media (100 μL) supplemented with the compounds to be screened at a 10 μM final concentration or a DMSO control (1 % DMSO final concentration in both cases).
6. Incubate the cells at 37 °C for 48 h.
7. Remove the media and rinse the cells with 1× PBS (50 μL).
8. Lyse the cells by adding 1× Passive Lysis Buffer (25 μL) and shake for 15 min at room temperature.
9. Assay the lysed cells using a Dual Luciferase Reporter Assay Kit according to the manufacturer's protocol and record the luminescence on a microplate reader. In short, for each well, dispense the Luciferase Assay Reagent II (100 μL), read firefly luciferase activity with a measurement time of 10 s and a delay time of 2 s, dispense the Stop & Glo Reagent (100 μL), read *Renilla* luciferase activity with a measurement time of 10 s and a delay time of 2 s.
10. Calculate the ratio of *Renilla* to firefly luciferase expression for each well to provide relative luciferase units (RLUs). The average and standard deviations for each of the triplicates is calculated from the RLU values.

3.3 Confirmation of Initial Hit Compounds by Dose Response

Compounds which elicit RLUs that are greater than the threshold $X + 3 \times \sigma$ (X = mean, σ = standard deviation) need to be re-assayed in triplicate to confirm the validity of the hit compound. In addition, Huh7-psiCHECK-miR122 stable cells should be treated with potential hit compounds in increasing concentration to test if a defined dose-dependent response can be established.

1. Passage Huh7-psiCHECK-miR122 cells into a white clear-bottom 96-well plate at 10,000 cells/well and grow overnight at 37 °C.
2. Replace the growth media with standard DMEM media (100 μL) supplemented with the small molecules (0.1, 0.5, 1, 5, and 10 μM final concentration) or a DMSO control (1 % DMSO final concentration) in triplicate.
3. Incubate the cells at 37 °C for 48 h.
4. Remove the media, wash the cells with 1× PBS, lyse the cells, and assay with a Dual Luciferase Reporter Assay Kit as previously described (*see* Subheading 3.2).
5. Calculate the ratio of *Renilla* to firefly luciferase expression for each well to provide relative luciferase units (RLUs). The average and standard deviations for each of the triplicates is calculated from the RLU values.
6. From the RLU data, generate a dose–response curve and calculate EC_{50} values, the concentration that gives 50 % of the compound's maximal response [22, 23].

3.4 Secondary psiCHECK-Control Assay

When using a luciferase-based reporter assay, it is possible for small molecules to increase the luciferase signal in a nonspecific fashion leading to false-positive hits [24, 25]. Therefore, putative hits which induce a significant increase in the relative luciferase signal should be assayed in Huh7 cells transfected with the psiCHECK-control plasmid to validate their activity as miR-122 inhibitors and confirm that they do not increase the luciferase signal in a non-miRNA-specific fashion. In the presence of the psiCHECK-control vector, there should be no statistically significant change in the normalized *Renilla* luciferase signal for the small molecule-treated cells.

1. Passage Huh7 cells into a white clear-bottom 96-well plate and grow overnight at 37 °C until they reach 80 % confluency.
2. Transfect the Huh7 cells with the psiCHECK-control plasmid (0.5 μg) using the X-tremGENE siRNA transfection reagent (3:2 reagent/DNA ratio) in Opti-Mem media (200 μL total volume) according to the manufacturer's protocol.
3. Incubate the cells at 37 °C for 4 h.

4. Replace the Opti-MEM media with standard DMEM growth media (100 μL) supplemented with the small molecules (10 μM) or a DMSO control (1 % DMSO final concentration) in triplicate.
5. Incubate the cells at 37 °C for 48 h.
6. Remove the media, wash the cells with 1× PBS, lyse the cells, and assay with a Dual Luciferase Reporter Assay Kit as described previously (*see* Subheading 3.2).
7. Calculate the ratio of *Renilla* to firefly luciferase expression for each well to provide relative luciferase units (RLUs). The average and standard deviations for each of the triplicates is calculated from the RLU values.

3.5 Secondary miR-21 Luciferase Assay

Compounds that increase luminescence in a miRNA-specific fashion could be specific inhibitors for miR-122 or could inhibit a general component of the miRNA pathway. In order to exclude general miRNA pathway inhibitors, the small molecules should be assayed with a luciferase reporter assay for a different miRNA. Here, a miRNA-21 reporter assay in Huh7 cells is used to validate the specificity of the miR-122 inhibitors, although other miRNAs could be used as control targets as well.

1. Passage Huh7 cells into a white clear-bottom 96-well plate and grow overnight at 37 °C until they reach 80 % confluency.
2. Transfect the Huh7 cells with the psiCHECK-miR21 plasmid (0.5 μg) using the X-tremGENE siRNA transfection reagent (3:2 reagent/DNA ratio) in Opti-Mem media (200 μL total volume) according to the manufacturer's protocol.
3. Incubate the cells at 37 °C for 4 h.
4. Replace the Opti-MEM media with standard DMEM growth media (100 μL) supplemented with the small molecules (10 μM final concentration) or a DMSO control (1 % DMSO final concentration) in triplicate.
5. Incubate the cells at 37 °C for 48 h.
6. Remove the media, wash the cells with 1× PBS, lyse the cells, and assay with a Dual Luciferase Reporter Assay Kit as described previously (*see* Subheading 3.2).
7. Calculate the ratio of *Renilla* to firefly luciferase expression for each well to provide relative luciferase units (RLUs). The average and standard deviations for each of the triplicates is calculated from the RLU values.

3.6 Quantitative RT PCR

The activity of the small molecule inhibitors can be further analyzed by quantitative Real-Time PCR (qRT PCR) to measure their direct effect on miR-122 expression levels in Huh7 cells.

The expression levels of a control miRNA (e.g., miR-21) should also be measured to further verify specificity for miR-122.

1. Passage Huh7 cells into a 6-well plate and grow overnight at 37 °C until they reach 60 % confluency.
2. Treat the cells with the small molecules (10 μM final concentration) or the DMSO control (1 % final DMSO concentration) in triplicate to ensure statistical validity.
3. Incubate the cells at 37 °C for 48 h.
4. Remove the media and wash the cells with 1× PBS buffer (2×2 mL).
5. Isolate the miRNA using a mirPremier microRNA Isolation Kit according to the manufacturer's protocol.
6. Quantify the miRNA using a Nanodrop ND-1000 spectrophotometer.
7. Perform a reverse transcription reaction with each RNA sample (10 ng) using a TaqMan microRNA Reverse Transcription Kit in conjunction with either the miR-122 or miR-21 TaqMan RT primer (16 °C, 30 min; 42 °C, 30 min; 85 °C, 5 min) according to the manufacturer's protocol.
8. Using the cDNA products, perform qRT PCR with the TaqMan Universal PCR Master Mix, No AmpErase UNG and the corresponding TaqMan miRNA assay probe on a Bio-Rad MyiQ Real-Time PCR Detection System (1.3 μL RT-PCR product; 95 °C, 10 min; followed by 40 cycles of 95 °C, 15 s; 60 °C, 60 s) according to the manufacturer's protocol.
9. The threshold cycles (C_t) can be used to determine the percent miRNA expression in small molecule-treated cells relative to the DMSO control, where DMSO represents 100 % expression.

4 Notes

1. The DMEM described here is made from DMEM/HIGH with L-glutamine powder (Hyclone). This growth media can also be purchased as pre-sterilized liquid media to simplify the media preparation. In this case, the pH does not need to be adjusted and only the FBS and penicillin/streptomycin need to be added.
2. Adjust the pH of the cell culture media before sterilization using HCl (6 M) or NaOH (2 M) made with ultrapure water.
3. TrypLE Express is a trypsin replacement that is gentle on cells, maintains healthy cell growth, and achieves faster dissociation. Trypsin EDTA, which is commonly used in cell culture laboratories, can also be used for passaging the Huh7 cell lines.

4. For the Dual Luciferase Reporter Assay, a microplate reader with dispensers is preferable for delivering the luciferase substrates; however, the procedure can be adjusted for microplate readers without dispensers (see manufacturer's protocol).

References

1. Tong AW, Nemunaitis J (2008) Modulation of miRNA activity in human cancer: a new paradigm for cancer gene therapy? Cancer Gene Ther 15:341–355
2. Sevignani C, Calin G, Siracusa L, Croce C (2006) Mammalian microRNAs: a small world for fine-tuning gene expression. Mamm Genome 17:189–202
3. Port JD, Sucharov C (2010) Role of microRNAs in cardiovascular disease: therapeutic challenges and potentials. J Cardiovasc Pharmacol 56:444–453
4. Lindsay MA (2008) microRNAs and the immune response. Trends Immunol 29:343–351
5. Cullen BR (2011) Viruses and microRNAs: RISCy interactions with serious consequences. Genes Dev 25:1881–1894
6. Deiters A (2010) Small molecule modifiers of the microRNA and RNA interference pathway. AAPS J 12:51–60
7. Meister G, Landthaler M, Dorsett Y, Tuschl T (2004) Sequence-specific inhibition of microRNA- and siRNA-induced RNA silencing. RNA 10:544–550
8. Krützfeldt J, Rajewsky N, Braich R, Rajeev K, Tuschl T, Manoharan M et al (2005) Silencing of microRNAs in vivo with "antagomirs". Nature 438:685–689
9. Ebert MS, Neilson JR, Sharp PA (2007) MicroRNA sponges: competitive inhibitors of small RNAs in mammalian cells. Nat Methods 4:721–726
10. Zhang S, Chen L, Jung E, Calin G (2010) Targeting microRNAs with small molecules: from dream to reality. Clin Pharmacol Ther 87:754–758
11. Garzon R, Marcucci G, Croce C (2010) Targeting microRNAs in cancer: rationale, strategies and challenges. Nat Rev Drug Discov 9:775–789
12. Shi XB, Tepper CG, deVere White RW (2008) Cancerous miRNAs and their regulation. Cell Cycle 7:1529–1538
13. Gumireddy K, Young D, Xiong X, Hogenesch J, Huang Q, Deiters A (2008) Small-molecule inhibitors of microrna miR-21 function. Angew Chem Int Ed Engl 47:7482–7484
14. Young D, Connelly C, Grohmann C, Deiters A (2010) Small molecule modifiers of microRNA miR-122 function for the treatment of hepatitis C virus infection and hepatocellular carcinoma. J Am Chem Soc 132:7976–7981
15. Connelly CM, Thomas M, Deiters A (2012) High-throughput luciferase reporter assay for small-molecule inhibitors of microRNA function. J Biomol Screen 17:822–828
16. Shan G, Li Y, Zhang J, Li W, Szulwach K, Duan R et al (2008) A small molecule enhances RNA interference and promotes microRNA processing. Nat Biotechnol 26:933–940
17. Watashi K, Yeung M, Starost M, Hosmane R, Jeang K (2010) Identification of small molecules that suppress microRNA function and reverse tumorigenesis. J Biol Chem 285: 24707–24716
18. Bose D, Jayaraj G, Suryawanshi H, Agarwala P, Pore SK, Banerjee R et al (2012) The tuberculosis drug streptomycin as a potential cancer therapeutic: inhibition of miR-21 function by directly targeting its precursor. Angew Chem Int Ed Engl 51:1019–1023
19. Chen X, Huang C, Zhang W, Wu Y, Chen X, Zhang C et al (2012) A universal activator of microRNAs identified from photoreaction products. Chem Commun 48:6432–6434
20. Jopling C, Yi M, Lancaster A, Lemon S, Sarnow P (2005) Modulation of hepatitis C virus RNA abundance by a liver-specific MicroRNA. Science 309:1577–1581
21. Lanford R, Hildebrandt-Eriksen E, Petri A, Persson R, Lindow M, Munk M et al (2010) Therapeutic silencing of microRNA-122 in primates with chronic hepatitis C virus infection. Science 327:198–201
22. Assay Guidance Manual Version 5.0, Eli Lilly and Company and NIH Chemical Genomics Center. Available online at: http://www.ncgc.nih.gov/guidance/manual_toc.html. Accessed Jan 2011
23. Sebaugh JL (2011) Guidelines for accurate EC50/IC50 estimation. Pharm Stat 10: 128–134
24. Thorne N, Inglese J, Auld DS (2010) Illuminating insights into firefly luciferase and other bioluminescent reporters used in chemical biology. Chem Biol 17:646–657
25. Auld DS, Thorne N, Nguyen DT, Inglese J (2008) A specific mechanism for nonspecific activation in reporter-gene assays. ACS Chem Biol 3:463–470

Chapter 13

Inhibition of miRNA Maturation by Peptide Nucleic Acids

Concetta Avitabile, Enrica Fabbri, Nicoletta Bianchi, Roberto Gambari, and Alessandra Romanelli

Abstract

Molecules able to interfere in miRNA genesis and function are potent tools to unravel maturation and processing pathways. Antisense oligonucleotides or analogs are actually employed for the inhibition of miRNA function. Here we illustrate how Peptide Nucleic Acids oligomers targeting pre-miRNA are exploited to inhibit miRNA maturation.

Key words Peptide nucleic acids, Pre-miR, miR, Peptide, FACS, RT-PCR

1 Introduction

Since the discovery of the small noncoding RNAs miRNAs as posttranscriptional regulators of gene expression, enormous efforts have been made to identify miRNA's biogenesis, targets and functions [1, 2]. MiRNAs are originated from precursors, pre-miRNAs. In humans a miR loading complex (miRLC) formed by Ago2 and Dicer has been identified as responsible for the miRNA maturation; pre-miRs are loaded and processed within this complex by the nuclease Dicer to generate miRNA duplexes [3, 4]. The duplexes are loaded onto the Ago proteins and successively unwinded; the guide strand is retained by Ago, while the passenger strand is degraded. In an alternative pathway, independent by Dicer, pre-miRs beginning with 5′U or 5′A associate to Ago2 in a miRNA precursor deposit complex (miDPC); Ago2 both cleaves the precursor and binds the mature miRNA [5]. MiRNAs are involved in several biological processes, including cell proliferation, differentiation and apoptosis; also several miRNAs have been demonstrated to be associated to pathological conditions [6]. In this context, the inhibition of miRNA function has become a priority task.

Here we describe a strategy for the inhibition of miRNA maturation using Peptide Nucleic Acid oligomers designed to target the

Christoph Arenz (ed.), *miRNA Maturation: Methods and Protocols*, Methods in Molecular Biology, vol. 1095, DOI 10.1007/978-1-62703-703-7_13, © Springer Science+Business Media New York 2014

miRNA precursor (pre-miRNA). Peptide Nucleic Acids were chosen as oligonucleotide analogs due to their high binding affinity and specificity toward RNA and high stability to degradation [7, 8]. Molecules able to bind to pre-miRNA affect the pre-miRNA processing and cause a reduction in the amount of mature miRNA produced and also pri-miR [9]. The case described refers to inhibition of pre-miR-210 maturation in human leukemia K562 cells, in which over-expression of miR-210 is induced in a time-dependent and dose-dependent fashion by mithramycin [10]. The strategy of pre-miR targeting is of general application, although it has to be considered that the protocols for the induction of miR expression depend on the cell system employed. The general criteria for the design of PNA anti-pre-miR are described. Protocols for the synthesis of the PNA-based molecules, evaluation of the PNA cellular uptake by FACS, quantification of pri-miR and mature miR are described.

The use of pre-miR targeting molecules has potential applications in the development of miRNA inhibitors in biomedicine and might also help discovering new details in the structural preferences and requirements necessary for the assembly of pre-miR processing complexes as miRLC and miDPC.

2 Materials

2.1 Synthesis (See Notes 1–3)

1. PAL-PEG resin LL (Applied Biosystems).
2. Dimethylformamide (DMF).
3. Dichloromethane (DCM).
4. Diethyl ether.
5. Fmoc-protected amino acids.
6. Fmoc-6-Ahx-OH.
7. 4-Methylmorpholine (NMM).
8. Pyridine.
9. Trifluoroacetic acid (TFA).
10. m-cresol.
11. Tri-isopropylsilane (TIS).
12. Fluorescein isothiocyanate isomer I (FITC).
13. Fmoc PNA monomers.
14. 0.22 M HOBT/HBTU solution: dissolve 166.9 mg of HBTU and 67.4 mg of HOBT in 2.0 mL of dry DMF. Vortex the solution until it is clear.
15. Capping solution: 15 % acetic anhydride, 15 % DIPEA, 70 % dry DMF (v/v).
16. Deblock solution: 30 % piperidine, 70 % dry DMF (v/v).

2.2 Cell Lines and Culture

1. K562 cells.
2. Tri-Reagent™.
3. RPMI 1640 medium supplemented with 10 % fetal bovine serum, 50 U/mL penicillin, and 50 U/mL streptomycin.
4. U6 snRNA.
5. PBS 1×.
6. 1 mM Mithramycin (MTH) stock solution: dissolve 1.085 mg of mithramycin in 1 mL of ultrapure water/methanol (1:1, v/v).

3 Methods

3.1 PNA Design

The PNA oligomer for the pre-miR targeting has to be perfectly complementary to the passenger strand of pre-miRNA, in the region which is complementary to the mature miRNA. Choose preferentially the region in which the passenger and the guide strand duplex show mismatches; this will help the local opening of the RNA duplex and the displacement of the complementary guide strand. The length of the PNA oligomer has to be chosen taking in consideration the base composition of the pre-miR and the stability of the duplex which will be formed between PNA and RNA. The PNA/RNA duplex has to be stable at the temperature at which the experiments will be carried out.

Always design also mismatched PNA oligomer, containing one or two mutations to verify the specificity of miR maturation inhibition.

Importantly, consider that the PNA oligomer will hardly enter into cells. Conjugation of PNA to carrier molecules is a requisite for a functional molecule. Peptides as TAT, the Nuclear Localization Signal (NLS), or polyArg efficiently deliver PNA into different type of cells [11, 12].

An example of PNA sequence designed for pre-miR-210 inhibition is reported in Fig. 1.

3.2 PNA-Peptide Synthesis by Fmoc-Chemistry

1. PNA-peptide conjugates are obtained by Fmoc solid phase synthesis on low loading resins as PAL-PEG-PS resin (0.18 mmol/g), on a 4 μmol scale following procedures reported in the literature [13, 14].
2. Synthesize first the peptide following standard procedures [14].
3. At the end of the synthesis remove the last Fmoc treating the resin with the Deblock solution. To verify whether the synthesis was successful, proceed from step 4 to 10 on a small aliquote of resin.
4. Wash the resin with DMF, DCM, and diethyl ether and finally dry the resin under vacuum.
5. Treat the dry resin with TFA/m-cresol/TIS 78/20/2 v/v/v solution 3 h. Use 1 mL solution for 10 mg of resin.
6. Transfer the TFA solution into an eppendorf tube.

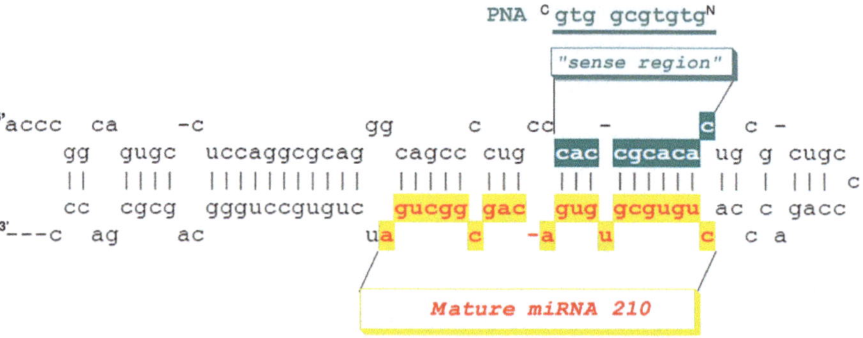

Fig. 1 Representation of the pre-miRNA-210 and the designed PNA anti-pre-miR (reproduced from ref. [9])

7. Evaporate TFA by nitrogen insufflation.
8. Add cold diethyl ether to precipitate the oligomer; keep the eppendorf in ice for 5 min; centrifuge 3 min at 11,000 rpm (8,000×*g*). Pour the ether in a separate tube and wash the precipitate again with ether.
9. Dissolve the crude in the minimum volume of H_2O and lyophilize it.
10. Analyze the crude by LC-MS.
11. Synthesize the PNA oligomer on the resin anchored peptide following standard procedures [13]. At the end of the synthesis cleave the conjugate off the resin, lyophilize the crude and HPLC purify the product (**steps 3–10**). Check the identity of the molecule by mass spectroscopy and its purity by HPLC.

3.3 Synthesis of the FITC-PNA-Peptide Conjugates

For the synthesis of the FITC derivative, use the PNA-peptide conjugates anchored to the resin, with the free N-terminal amino group (*see* **Note 4**).

1. Swell the resin in DMF for 1 h.
2. Dissolve 7.06 mg of Fmoc-ε–Ahx-OH in a 89.1 μL of 0.22 M solution of HOBT/HBTU in DMF and add 3.08 μL of NMM. Add the solution to the resin and let the reaction proceed for 40 min.
3. Wash the resin with DMF.
4. Add 0.5 mL of capping solution and let the reaction proceed for 5 min.
5. Wash extensively the resin with DMF.
6. Treat the resin with the deblock solution for 5 min (2 times).
7. Protect the reaction vessel from the light (*see* **Note 3**).
8. In a dark tube dissolve 15.56 mg of FITC in 200 μL of DMF. Add 6.17 μL of NMM. Transfer the solution to the reaction

vessel and let the reaction proceed for 1 h. Repeat the FITC coupling.

9. Repeat Steps from Subheading 2.2, **step 4** to Subheading 3.2, **step 10**.
10. HPLC purify the conjugate.
11. Check the identity of the molecule by mass spectroscopy and its purity by HPLC.

3.4 Human Cell Lines and Culture Conditions

1. Culture human leukemia K562 cells in humidified atmosphere of 5 % CO_2/air in RPMI 1640 medium supplemented with 10 % fetal bovine serum, 50 U/mL penicillin, and 50 μg/mL streptomycin [15].
2. Prepare a stock solution of Mithramycin (MTH) 100 μM; store the solution at −20 °C in the dark and dilute it immediately before the use [15, 16].
3. Treat the K562 cells with 15 nM of Mithramycin (MTH) at the beginning of the cultures (seed 30,000 cells /mL).

3.5 Uptake Evaluation by FACS

1. Dissolve the lyophilized PNA in ultrapure water at a 500 μM concentration.
2. Incubate the cells 3×10^4 cell/mL (K562 previously treated with mithramycin) with and without 2 μM of the PNA-peptide-FITC conjugate for 48 h.
3. Harvest and wash the cells.
4. Analyze 1×10^5 cells by the CellQuestTM version 3.3 software (Becton Dickinson), using the FL1 channel to detect fluorescence.
5. Repeat the experiment at least 3 times for statistical analysis.
6. Express the results as median fold, i.e., the ratio between the median fluorescence intensity values obtained in the presence and absence of treatment, respectively. Data can be presented with an histogram, showing the number of cells *versus* the expressed fluorescence intensity.

An example of FACS data is reported in Fig. 2 for K562 cells treated with FITC-PNA-peptide conjugates.

3.6 RNA Extraction

1. Isolate cells by centrifugation at 1,500 rpm (1,600 × *g*) for 10 min at 4 °C and wash in PBS 1×.
2. Lyse the cells with Tri-Reagent™, according to manufacturer's instructions.
3. Precipitate the isolated RNA in 75 % cold ethanol and store it at −80 °C.
4. Dry and dissolve RNA in ultrapure nuclease-free water before use.

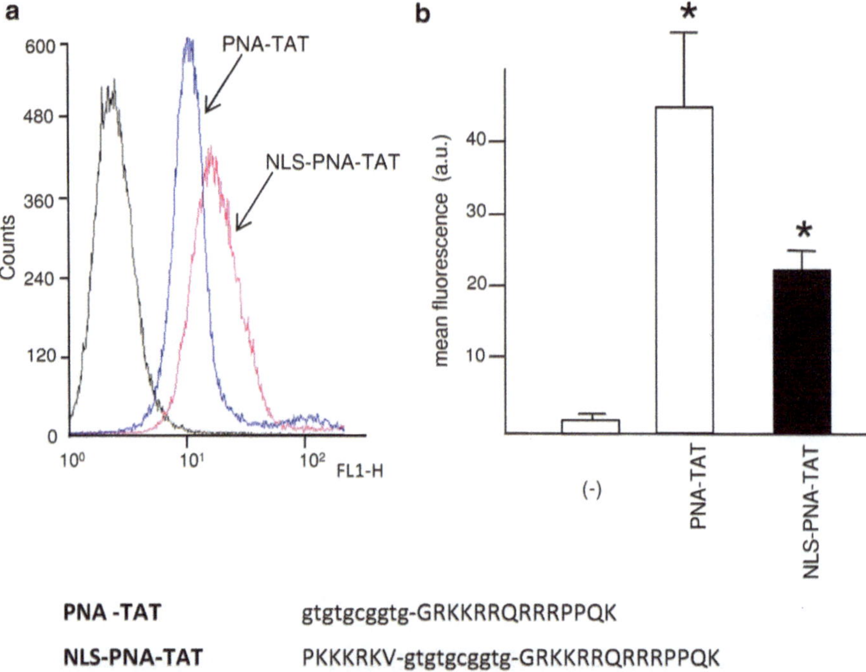

Fig. 2 FACS analysis showing the uptake of PNA-TAT and NLS-PNA-TAT after 48 h incubation of K562 cells in the absence (*minus*) or in the presence of 2 μM concentrations of the fluorescein-labeled PNA-TAT ((**a**), *pink line*), and NLS-PNA-TAT ((**a**), *blue line*) molecules. In (**b**) the quantitative determinations are shown. Sequences of the PNA and peptides are also shown in the lower part of the panel. *Lower case letters* are employed for PNA bases, *upper case letters* for amino acids

3.7 Pri-miRNA and miRNA Quantification

1. Isolate RNA from K562 cells and measure it by reverse transcription quantitative real-time polymerase chain reaction (qRT-PCR) using gene-specific double fluorescence labeled probes. In the example reported an ABI Prism 7700 Sequence Detection System version 1.7.3 (Applied Biosystems, Monza, Italy) was employed, according to the manufacturer's protocols.
2. Use for each sample 20 ng of cDNA for the assays. All RT reactions, including no-template controls and RT-minus controls. Perform the experiment in duplicate.
3. Calculate the relative expression using the comparative cycle threshold method and as reference U6 snRNA to normalize all RNA samples, since it remains constant in the assayed samples by miR-profiling and quantitative RT-PCR analysis.

An example of pri-miR and miR quantification is reported in Fig. 3, using PNA-peptide conjugates designed against pre-miR-210.

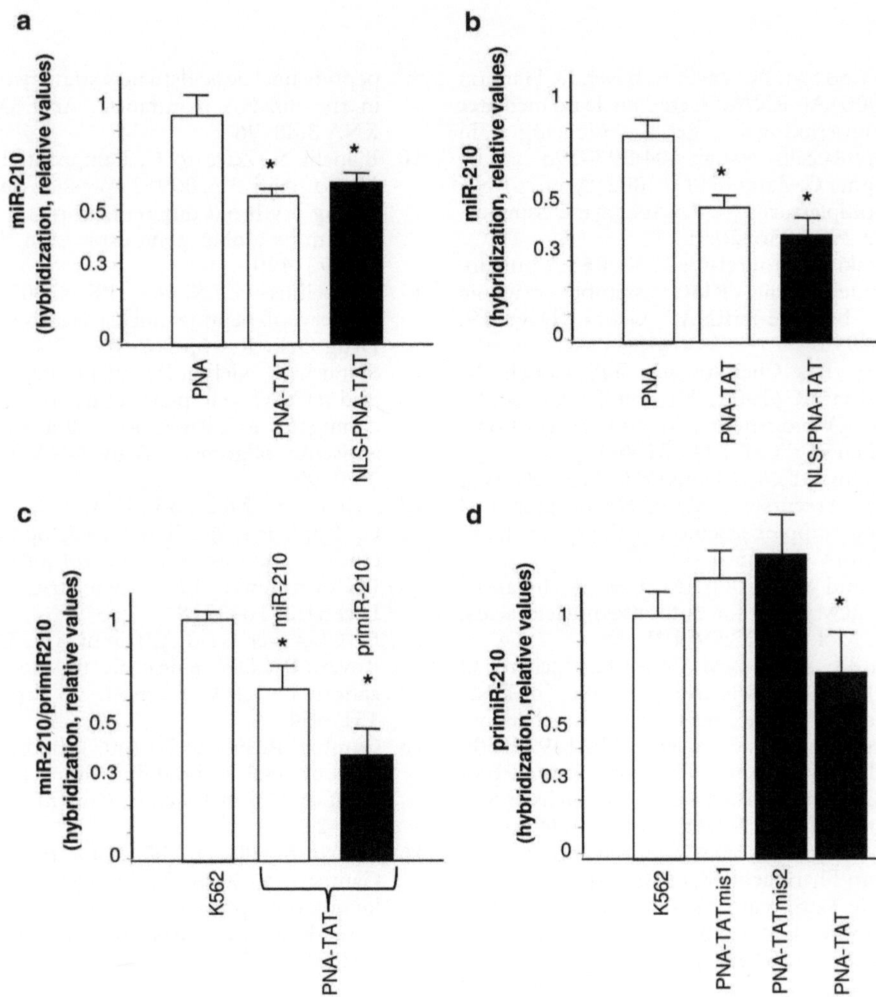

Fig. 3 (**a, b**) Quantification by real-time PCR of miRNA-210 in Mithramycin-induced K562 cells. Effects of treatment with PNA-TAT and NLS-PNA-TAT, 1 μM (**a**) and 2 μM (**b**) on miRNA-210 content in cells cultured for 48 h in the presence of MTH 15nM. (**c**) Effects on miR-210 and pri-miR-210 after 72 h treatment of K562 cells with 2 μM PNA-TAT. (**d**) Effects of 24 h treatment with 1 μM PNA-TAT-mis1, PNA-TATmis2 (control sequences containing 1 mismatch as compared to PNA-TAT), and PNA-TAT on pri-miR-210. The data represent the average ± SD ($n = 3$) (*$p < 0.05$) (reproduced from ref. [9])

4 Notes

1. Carry out all synthetic protocols under the fume hood, wearing appropriate protections (gloves and glasses).
2. Always prepare fresh solutions for activators, capping and deblock.
3. Protect the capping and deblock solutions from the light.
4. Steps from Subheading 3.3, **step 8** to Subheading 3.3, **step 11** and all work with FITC-modified oligomers need to be always carried out in the dark.

References

1. Hammond SM, Bernstein E, Beach D, Hannon GJ (2000) An RNA-directed nuclease mediates post-transcriptional gene silencing in Drosophila cells. Nature 404:293–296
2. Hutvagner G, Zamore PD (2002) A microRNA in a multiple-turnover RNAi enzyme complex. Science 297:2056–2060
3. Maniataki E, Mourelatos Z (2005) A human, ATP-independent, RISC assembly machine fueled by pre-miRNA. Genes Dev 19: 2979–2990
4. Gregory RI, Chendrimada TP, Cooch N, Shiekhattar R (2005) Human RISC couples microRNA biogenesis and posttranscriptional gene silencing. Cell 123:631–640
5. Liu X, Jin DY, McManus MT, Mourelatos Z (2012) Precursor microRNA-programmed silencing complex assembly pathways in mammals. Mol Cell 46:507–517
6. Hammond SM (2006) MicroRNA therapeutics: a new niche for antisense nucleic acids. Trends Mol Med 12:99–101
7. Nielsen PE, Egholm M, Berg RH, Buchardt O (1991) Sequence-selective recognition of DNA by strand displacement with a thymine-substituted polyamide. Science 254:1497–1500
8. Demidov VV, Potaman VN, Frank-Kamenetskii MD, Egholm M, Buchard O, Sonnichsen SH, Nielsen PE (1994) Stability of peptide nucleic acids in human serum and cellular extracts. Biochem Pharmacol 48:1310–1313
9. Avitabile C, Saviano M, D'Andrea L, Bianchi N, Fabbri E, Brognara E, Gambari R, Romanelli A (2012) Targeting pre-miRNA by peptide nucleic acids: a new strategy to interfere in the miRNA maturation. Artif DNA PNA XNA 3:88–96
10. Bianchi N, Zuccato C, Lampronti I, Borgatti M, Gambari R (2009) Expression of miR-210 during erythroid differentiation and induction of gamma-globin gene expression. BMB Rep 42:493–499
11. Koppelhus U, Nielsen PE (2003) Cellular delivery of peptide nucleic acid (PNA). Adv Drug Deliv Rev 55:267–280
12. Shiraishi T, Nielsen PE (2011) Peptide nucleic acid (PNA) cell penetrating peptide (CPP) conjugates as carriers for cellular delivery of antisense oligomers. Artif DNA PNA XNA 2:90–99
13. Avitabile C, Moggio L, D'Andrea LD, Pedone C, Romanelli A (2010) Development of an efficient and low-cost protocol for the manual PNA synthesis by Fmoc chemistry. Tetrahedron Lett 51:3716–3718
14. Le Chevalier Isaad A, Papini AM, Chorev M, Rovero P (2009) Side chain-to-side chain cyclization by click reaction. J Pept Sci 15: 451–454
15. Gambari R, Fibach E (2007) Medicinal chemistry of fetal hemoglobin inducers for treatment of beta-thalassemia. Curr Med Chem 14: 199–212
16. Fibach E, Bianchi N, Borgatti M, Prus E, Gambari R (2003) Mithramycin induces fetal hemoglobin production in normal and thalassemic human erythroid precursor cells. Blood 102:1276–1281

Chapter 14

Molecular Methods for Validation of the Biological Activity of Peptide Nucleic Acids Targeting MicroRNAs

Eleonora Brognara, Enrica Fabbri, Nicoletta Bianchi, Alessia Finotti, Roberto Corradini, and Roberto Gambari

Abstract

The involvement of microRNAs in human pathologies is a firmly established fact. Accordingly, the pharmacological modulation of their activity appears to be a very appealing issue in the development of new types of drugs (miRNA therapeutics). One of the most interesting issues is the possible development of miRNA therapeutics for development of anti-cancer molecules. In this respect appealing molecules are based on peptide nucleic acids (PNAs), displaying a pseudo-peptide backbone composed of *N*-(2-aminoethyl)glycine units and found to be excellent candidates for antisense and antigéne therapies. The major limit in the use of PNAs for alteration of gene expression is the low uptake by eukaryotic cells. The aim of this chapter is to describe methods for determining the activity of PNAs designed to oncomiRNA targets, using as model system miR-221 and its target $p27^{Kip1}$ mRNA. The effects of PNAs targeting miR-221 are here presented discussing data obtained using as model system the human breast cancer cell line MDA-MB-231, in which miR-221 is up-regulated and $p27^{Kip1}$ down-regulated.

Key words PNA, MicroRNA, Cell assays, Anti-miRs, FACS analysis, Taqman assays

Abbreviation

AEEA	2-(2-aminoethoxy)ethoxyacetyl spacer
FACS	Fluorescence-activated cell sorter
FBS	Fetal bovine serum
Fl	Fluorescein
PBS	Phosphate-buffered saline
PNA	Peptide nucleic acid
3′UTR	3′-untranslated region
RT-qPCR	Retro transcription-quantitative polymerase chain reaction
EDTA	Ethylenediaminetetraacetic acid
SDS	Sodium dodecyl sulfate
DTT	Dithiotreithol
TBS	Tris-buffered saline

Christoph Arenz (ed.), *miRNA Maturation: Methods and Protocols*, Methods in Molecular Biology, vol. 1095, DOI 10.1007/978-1-62703-703-7_14, © Springer Science+Business Media New York 2014

HRP Horseradish peroxidase
RISC RNA-induced silencing complex

1 Introduction

MicroRNAs (miRNAs, miRs) are a family of small (19–25 nucleotides in length) noncoding RNAs that regulate gene expression by sequence-selective targeting of mRNAs [1], leading to a translational repression or mRNA degradation, depending on the degree of complementarity between miRNAs and the target sequences [2]. Since a single miRNA can target several mRNAs and a single mRNA may contain several signals for miRNA recognition, it is calculated that at least 10–40 % of human mRNAs are targets of microRNAs [1]. In general, a low expression of a given miRNA is expected to be potentially linked with an accumulation of targets mRNAs; conversely, a high expression of miRNAs is expected to be the cause of a low expression of the target mRNAs [2, 3].

Since the involvement of microRNAs in human pathologies is a firmly established fact, the pharmacological modulation of their activity appears to be a very appealing issue in the development of new types of drugs (miRNA therapeutics). One of the most interesting issues is the possible development of miRNA therapeutics for development of anti-cancer molecules [4]. In this respect peptide nucleic acid (PNA)-based molecules are appealing [5].

In PNAs the pseudo-peptide backbone is composed of *N*-(2-aminoethyl)glycine units [6]. PNAs are resistant to both nucleases and proteases [7, 8] and, more importantly, hybridize with high affinity to complementary sequences of single-stranded RNA and DNA, forming Watson-Crick double helices [1]. For these reasons, PNAs were found to be excellent candidates for antisense and antigéne therapies [9–11]. The major limit in the use of PNA for alteration of gene expression is the low uptake by eukaryotic cells [12]. In order to solve this drawback, several approaches have been considered, including the delivery of PNA analogues with liposomes and microspheres [7, 13, 14]. One of the possible strategy is to link PNAs to polyarginine (R) tails, based on the observation that this cell-membrane penetrating oligopeptides are able to facilitate uptake of conjugated molecules [15]. Peptide–PNA conjugates have been shown to be efficiently incorporated in cells by gymnosis, i.e., without the need of transfecting agents, showing high uptake efficiency [16].

The aim of this chapter is to describe methods for determining the activity of PNAs designed to target oncomiRNAs, using as model system miR-221 and its target $p27^{Kip1}$ mRNA [17–19]. The effects of PNAs targeting miR-221 are here presented discussing data obtained using as cellular model system the human breast cancer cell line MDA-MB-231, in which miR-221 is up-regulated and $p27^{Kip1}$ down-regulated [20].

2 Materials

2.1 Peptide Nucleic Acids (PNAs)

PNAs can be purchased from several companies, including Panagene, Inc. (www.panagene.com). Alternatively, PNAs against miRNAs can be synthesized following the procedures described in Manicardi et al. [2] and Fabbri et al. [3]. The data here presented are based on PNAs described in the paper by Brognara et al. [22]. The sequences of PNA-a221 and Rpep-PNA-a221 are reported in Table 1; fluorescein-labeled PNAs were used for the uptake experiments.

2.2 Cell Culture

1. RPMI-1640 medium supplemented with 10 % fetal bovine serum, 50 units/ml penicillin, and 50 mg/ml streptomycin.
2. LHC-8 basal medium without gentamycin.
3. D-MEM medium.
4. 2 mM L-Glutamine.
5. Human erythroleukemic K562 cells.
6. Bronchial epithelial IB3-1 cell line, derived from Cystic Fibrosis patients.
7. Human breast cancer cell lines MCF-7 and MDA-MB-231.
8. To determine the effects on proliferation, cell growth is monitored using a Z1 Coulter Counter (Coulter Electronics, Hialeah, FL, USA).
9. Cell incubation and monitoring system like BioStation instrument (BioStationIM, Nikon Instruments).

2.3 FACS Analysis

1. Trypsin–EDTA.
2. Fetal bovine serum, FBS.
3. Phosphate-buffered saline, PBS.
4. Determination of cellular uptake: fluorescence intensity was detected using the FACScan (Becton Dickinson, Franklin Lakes, NJ, USA). Cells were analyzed using the CellQuest™ version 3.3 software (Becton Dickinson, Franklin Lakes, NJ, USA).

Table 1
Sequences of PNAs

PNA-a221	H-AAACCCAGCAGACAATGT-NH_2
Fl-PNA-a221	Fl-AEEA-AAACCCAGCAGACAATGT-NH_2
Rpep-PNA-a221	H-RRRRRRRR-AAACCCAGCAGACAATGT-NH_2
Fl-Rpep-PNA-a221	Fl-AEEA-RRRRRRRR-AAACCCAGCAGAC AATGT-NH_2

Fl fluorescein; *AEEA* 2-(2-aminoethoxy)ethoxyacetyl spacer

2.4 RNA Extraction

1. Trypsin–EDTA.
2. Fetal bovine serum, FBS.
3. Phosphate-buffered saline, PBS.
4. Tri-Reagent™, for cell lysis.

2.5 Real-Time Quantitative PCR of MicroRNA

1. Primers and probes can been obtained from local supplier, like Applied Biosystems.
2. Reverse transcriptase (RT) reactions are performed using the TaqMan® MicroRNA Reverse Transcription Kit (Applied Biosystems, Foster City, CA, USA).
3. Real-time PCR machine, like the 7700 Sequence Detection System version 1.7 (Applied Biosystems, Foster City, CA, USA).

2.6 Taqman RT-qPCR for Gene Expression Analysis

1. Primers and probes to assay p27^{Kip1} (Assay ID: Hs00153277.m1) and the endogenous control human 18S rRNA can be purchased from Applied Biosystems (Applied Biosystems, Foster City, CA, USA).
2. Reverse transcriptase (RT) reactions are preferentially performed using the TaqMan® Reverse Transcription Kit (Applied Biosystems, Foster City, CA, USA). RNAs are reverse transcribed by using random hexamers.

2.7 Western Blotting

1. SDS-denaturing buffer (1×): 62,5 mM Tris–HCl (pH 6.8), 2 % SDS, 50 mM DTT, 0.01 % bromophenol blue, 10 % glycerol for cytoplasmic extracts.
2. SDS-PAGE running buffer: 25 mM Tris–HCl, pH 8.3, 192 mM glycine, 0.1 % SDS.
3. Biotinylated protein ladder (size range of 9–200 kDa).
4. Nitrocellulose membrane: 20 μm nitrocellulose membrane.
5. Western blot transfer buffer: 25 mM Tris, 192 mM Glycine, 5 % methanol.
6. Pre-staining Solution: Ponceau S Solution.
7. Tris-buffered saline, TBS: 10 mM Tris–HCl (pH 7.4), 150 mM NaCl.
8. TBS containing 0.1 % Tween-20 (TBS/T).
9. Blocking solution: 5 % nonfat milk in TBS/T buffer. Store at 4 °C.
10. Primary monoclonal antibodies.
11. HRP-conjugated secondary antibody.
12. Chemoluminescent substrate like LumiGLO® (0.5 ml 20× LumiGLO®, 0.5 ml 20× Peroxide, and 9.0 ml Milli-Q water), Cell Signaling.

13. X-ray film.
14. Western Blot Stripping Buffer, like Restore™ (Pierce).
15. Film analyzer like Gel Doc 2000 (Bio-Rad).

2.8 Statistics

Results are expressed as mean ± standard error of the mean (SEM). Comparisons between groups are done by using paired Student's *t*-test and a one-way analysis of variance (ANOVA). Statistical significance is defined with $p < 0.01$.

3 Methods

3.1 Human Cell Lines, PNA Treatment, and Culture Conditions

1. Incubate all the cell lines in humidified atmosphere of 5 % CO_2/air at 37 °C. Human erythroleukemic K562 cells are maintained in suspension culture using RPMI-1640 medium. Bronchial epithelial IB3-1 cell line, derived from Cystic Fibrosis patients, were grown in LHC-8 basal medium, supplemented with 5 % FBS in the absence of gentamycin. Incubate human breast cancer MCF-7 and MDA-MB-231 in D-MEM medium supplemented with 10 % fetal bovine serum and 2 mM L-Glutamine. Determine the effects on proliferation, monitor cell growth by determining the cell number/ml using a Z1 Coulter Counter.
2. Add increasing concentrations (from 500 nM to 10 μM) of PNAs targeting miR-221 to cell cultures. Seed K562 at initial concentration of 1×10^4 cells/ml; determine cell growth, according to cell number/ml, usually after 3, 4, and 5 days of treatment, using the Z1 Cell Counter. Seed IB3-1 and MDA-MB-231 cells at the initial concentration of 3×10^4 cells/cm^2; seed MCF-7 at the initial concentration of 1×10^5 cells/cm^2 and determine the cell number/ml after 3 days of treatment. Determine Cell number/ml after trypsin treatment by using the Z1 Cell Counter.

3.2 Uptake of Anti-miR-221 PNAs

1. Incubate, e.g., 1×10^5 K562 cells in the presence of increasing concentrations of fluorescein-labeled PNAs (Fl-PNA-a221 and Fl-Rpep-PNA-a221) in order to investigate the uptake of PNA-a221 and Rpep-PNA-a221 (*see* **Notes 1** and **2**).
2. Perform FACS analyses to obtaining results like shown in Fig. 1 after 12 (Fig. 1a), 24 (Fig. 1b), and 48 (Fig. 1c) hours of culture. Usually after 12 h of treatment, low binding of the Fl-PNA-a221 to target K562 cells is observed. On the contrary, Fl-Rpep-PNA-a221 will efficiently bind to K562 cells (Fig. 1a).
3. Analyze the uptake of Fl-PNA-a221 and Fl-Rpep-PNA-a221 by using the BioStation instrument. Representative results are shown in Fig. 1 (D–F) and indicate the differential uptake (fully consistent with the FACS analysis) obtained using

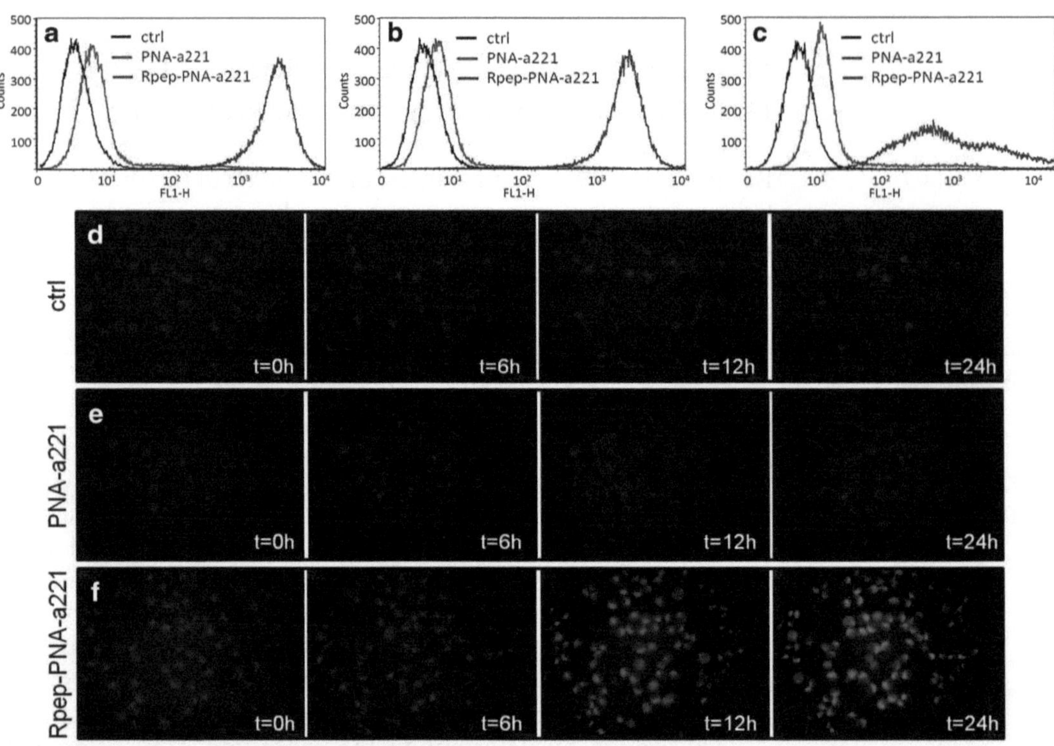

Fig. 1 FACS analysis after 12 h (**a**), 24 h (**b**), and 48 h (**c**) incubation of K562 cells in the absence (ctrl) or in the presence of 2 μM of the fluorescein labeled Fl-PNA-a221 (*red lines*) and Fl-Rpep-PNA-a221 (*blue lines*) molecules. (**d**–**f**) Intracellular distribution of Fl-PNAs. In this experiment, untreated cells (**d**) and cells treated with 2 μM Fl-PNA-a221 (**e**) and Fl-Rpep-PNA-a221 (**f**) were analyzed using the Nikon BioStation IM, an integrated cell incubator and fluorescence camera system, 20i, magnification, ×20. Pictures of cell culture were taken at the start of treatment (t = 0) and after 6, 12, and 24 h, as indicated

Fl-PNA-a221 and Fl-Rpep-PNA-a221 fluorescent compounds. The highest uptake is obtained when fluorescein-labeled Fl-Rpep-PNA-a221 is used (Fig. 1f). A low level of fluorescence is detectable with PNA-a221 (Fig. 1e).

4. Confirm uptake by FACS analysis after 24 h of treatment (Fig. 2) on human erythroleukemic K562 (Fig. 2a), bronchial epithelial IB3-1 (Fig. 2b), breast cancer MCF-7 (Fig. 2c), and MDA-MB-231 (Fig. 2d) cells: a good uptake can be obtained in all the cell lines only by Fl-Rpep-PNA-a221 (*see* **Notes 3** and **4**).

3.3 RNA Extraction and Real-Time Quantitative Analysis of miRNA Expression

1. Trypsinize cells and collect by centrifugation at 1,600 × *g* for 10 min at 4 °C, wash with PBS, lyse with Tri-Reagent™, according to manufacturer's instructions. Wash the isolated RNA once with cold 75 % ethanol, dry and dissolve in nuclease-free pure water before use.
2. For microRNA quantification perform reverse transcriptase (RT) reactions using the TaqMan® MicroRNA Reverse

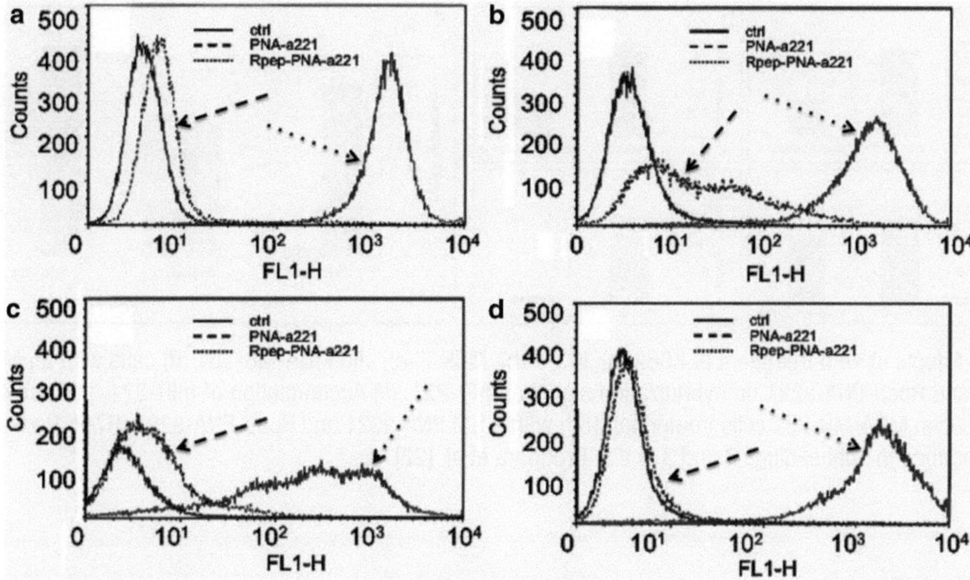

Fig. 2 FACS analysis performed after 24 h of culturing K562 (**a**), IB3-1 (**b**), MCF-7 (**c**), and MDA-MB-231 (**d**) cells in the absence (ctrl) or in the presence of 2 μM fluorescein labeled Fl-PNA-a221 (*dashed arrows*) and Fl-Rpep-PNA-a221 (*dotted arrows*) molecules

Transcription Kit; perform real-time PCR according to the manufacturer's protocols. Use 20 ng per sample for the assays. Perform all RT reactions, including no-template controls and RT-minus controls, in duplicate using the 7700 Sequence Detection System version 1.7. Calculate the relative expression using the comparative cycle threshold method and use U6 snRNA as reference to normalize all RNA samples, since it remains constant in the assayed samples by miR-profiling and quantitative RT-PCR analysis, as reported previously [3, 21].

3. When K562, IB3-1, MCF-7, and MDA-MB-231 cells are cultured in the presence of PNA-a221 and Rpep-PNA-a221 similar effects were observed on miRNA-221. After RNA isolation, perform RT-qPCR following protocols like described in the chapter by Zoellner et al. and apply to analysis of miR-221 [3]. Figure 3a–d demonstrate that the miR-221-specific hybridization signal is strongly reduced only when RNA is isolated from cells cultured for 48 h in the presence of Rpep-PNA-a221, while no major effects are observed for PNA-a221 (*see* **Note 5**).
4. Figure 3e shows that these effects are restricted to miR-221, since, despite the fact that some alteration of miRNA content occurs, no suppression of accumulation of miR-200c and miR-let-7c has been obtained. These data demonstrate specificity of PNA-mediated effects.

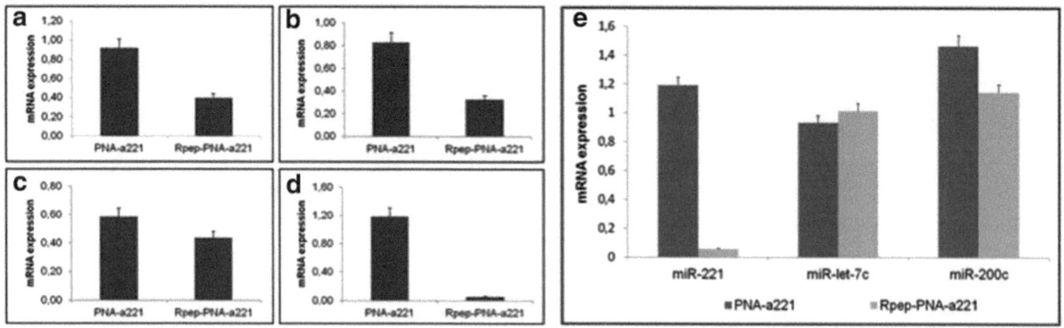

Fig. 3 Effects of 48 h treatment of K562 (**a**), IB3-1 (**b**), MCF-7 (**c**), and MDA-MB-231 (**d**) cells with 2 μM PNA-a221 and Rpep-PNA-a221 on hybridization signal of miR-221. (**e**) Accumulation of miR-221, miR-let-7c and miR200c in MDA-MB-231 cells treated for 48 h with 2 μM PNA-a221 and Rpep-PNA-a221. RT-PCR was done as described in Subheadings 2 and 3 and in Brognara et al. [22]

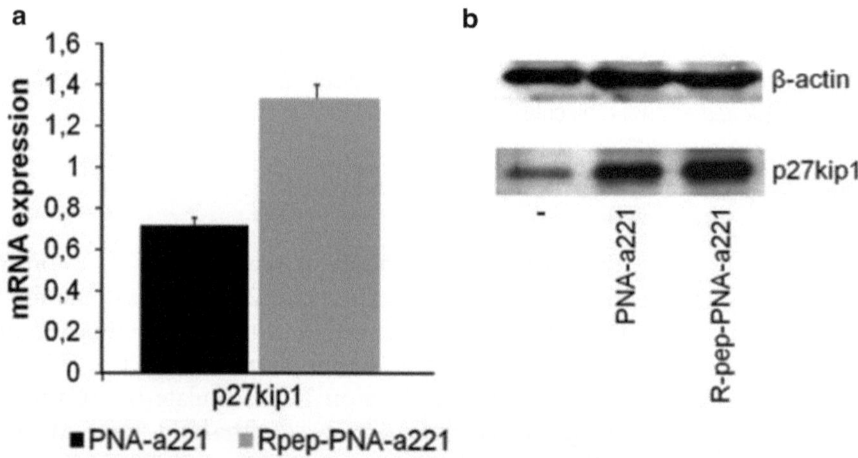

Fig. 4 Accumulation of p27[Kip1] mRNA (**a**) in MDA-MB-231 cells treated for 96 h with 2 μM PNA-a221 and Rpep-PNA-a221. (**b**) Western blotting performed on the same cellular populations using antibody against p27[Kip1] and against β-actin as reference protein. Modified from Brognara et al. [22]

3.4 Studies on Alteration of Gene Expression

1. For gene expression analysis (*see* **Note 6**) reverse transcribe 500 ng of the total RNA by using random hexamers. Carry out quantitative real-time PCR assays using gene-specific double fluorescently labeled probes. The relative expression is calculated using the comparative cycle threshold method and, as reference genes, the endogenous control human 18S rRNA [21].
2. RT-qPCR analyses. The effects on p27[Kip1] mRNA, shown in Fig. 4a, indicate that no change of p27[Kip1] mRNA content occurs in MDA-MB-231 cells in the presence of PNA-a221, whereas significant increase of p27[Kip1] mRNA is observed with the Rpep-PNA-a221 ($p < 0.05$).

3. Western blotting. Denature 20 µg of cytoplasmic extracts for 5 min at 98 °C in 1× SDS sample buffer and load on SDS-PAGE gel (10×8 cm) in Tris-glycine Buffer. Use a biotinylated protein ladder (size range of 9–200 kDa) as standard to determine molecular weight. The electrotransfer to 20 µm nitrocellulose membrane is performed overnight at 360 mA and 4 °C in electrotransfer buffer. The membrane is prestained in Ponceau S Solution to verify the transfer, washed with 25 ml TBS for 10 min at room temperature and incubated in 25 ml of blocking buffer for 2 h at room temperature. The membranes are washed three times for 5 min each with 25 ml of TBS/T and incubated with primary rabbit monoclonal antibody (1:1,000) in 15 ml primary antibody dilution buffer with gentle agitation overnight at 4 °C. The day after, the membrane are washed three times for 5 min each with 20 ml of TBS/T and incubated in 15 ml of blocking buffer, in gentle agitation for 2 h at room temperature, with an appropriate HRP-conjugated secondary antibody (1:2,000) and an HRP-conjugated anti-biotin antibody (1:1,000) to detect biotinylated protein marker. Finally, after three washes each with 20 ml of TBS/T for 5 min, the membranes are incubated with chemoluminescent substrate according to the manufacturers instruction and exposed to X-ray film. X-ray films for chemiluminescent blots are analyzed by Gel Doc 2000 using Quantity One program to elaborate the intensity data of your specific protein targets. Ponceau S staining can be used as normalization control, but others marker proteins may be taken as reference too and are specifically reported. The data of Western blot assay (Fig. 4b) show a clear increment of p27^{Kip1} protein expression in the sample from cells treated with Rpep-PNA-a221.

4 Notes

1. FACS analyses, such as those shown in Fig. 1 (panels A–C) even performed several times and obtaining consistent results, are compatible with uptake of Fl-Rpep-PNA-a221 by target cells, but cannot exclude the possibility that the fluorescence signal is due at least partially to cell–surface interactions, caused by the positive charged polyarginine peptide, which might interact strongly with negative charged protein components. Therefore FACS-based assays should be combined with microscope-assisted analyses (*see* Fig. 1, panels D–F) to confirm that PNAs are internalized within target cells. Confocal analyses are in addition necessary if compartmentalization within intracellular compartments should be determined.
2. In order to determine the concentrations of Fl-Rpep, Fl-PNA-a221, and Fl-Rpep-PNA-a221 to be employed for in

vitro studies on target cells, the IC_{50} after treatment for different days should be always determined. In our case, while Fl-PNA-a221 displayed on MDA-MB-231 an IC_{50} value higher that 15 μM, the IC_{50} value of Fl-Rpep-PNA-a221 was found to be 7.5 ± 1.75 μM. Accordingly, in order to avoid to use antiproliferative (and possibly cytotoxic) concentrations, Fl-Rpep, Fl-PNA-a221, and Fl-Rpep-PNA-a221 were used at 2 μM.

3. Unlike commercially avalilable anti-miR molecules, most of which need continuous administrations, a single administration of anti-miR PNAs is sufficient to obtain the biological effects on microRNA activity. In any case, modifications allowing efficient uptake by target cells are necessary to obtain the biological activity; it should be noted that in our case PNA-a221, despite being able to hybridize to the target nucleotide sequence, is not internalized (Figs. 1 and 2) and displays a very low activity on MDA-MB-231 breast cancer cells (Figs. 3 and 4).

4. Conjugation with the octaarginine peptide, according to a previously developed strategy for K562 cells, is effective not only on the MDA-MB-231 cellular system but also on other cell lines (*see* Fig. 2). We would like to underline that the delivery of Rpep-PNA-a221 needs no transfection reagents (i.e., lipofectin, lipofectamine, and similar reagents) which, on the contrary, are required when RNA- or DNA-based analogues are used. The octaarginine peptide is expected to have also beneficial effects in terms of affinity for nucleic acid targets, by an additional contribution of electrostatic interactions to the sequence specific base-pairing interactions of the PNA [22].

5. Explanation of the effects of PNAs on microRNA accumulation. PNAs might interact very stably with target mRNAs. Therefore, before performing RT-PCR, it is necessary to demonstrate that PNAs are not co-purified with target mRNAs. High-quality RNA preparation are necessary. In addition, the effects of PNA addition on RT reaction and on PCR efficiency should be analyzed. Complementary analyses (for instance Northern blotting) might be also considered using oligonucleotide probes not affected in their hybridization efficiency by pre-formed PNA–RNA hybrids. In studies similar to these described in this paper, this issue is the most critical. Studies suggesting a PNA-mediated effect on microRNA content should be in any case accompanied by studies on the effects on the expression of mRNAs demonstrated to be target of the studied microRNAs.

6. As anticipated in **Note 5**, the effects of PNA against microRNAs should be analyzed on the expression of relevant target mRNAs. In our example, several mRNAs have been firmly established to be target molecules of miR-221, such as DVL2, PUMA, PTEN, $p27^{Kip1}$ [22]. In the context of breast tumors,

of great interest is the study published by Farace et al. identifying p27^{Kip1} mRNA as a possible target of miR-221 [19]. This finding is very intriguing, since p27^{Kip1} has been proposed as a tumor-suppressor gene, which is down-regulated in several types of tumors. These data support the concept that targeting miR-221 with antagomiR molecules might lead to an increased expression of the tumor-suppressor p27^{Kip1}, bringing novel treatment options in cancer treatment.

Acknowledgments

This work was partially supported by a grant from MIUR (PRIN09 grant n. 20093N774P "Molecular recognition of microRNA (miR) by modified PNA: from structure to activity"). R.G. is granted by Fondazione Cariparo (Cassa di Risparmio di Padova e Rovigo), by UE THALAMOSS (THALAssaemia MOdular Stratification System for personalized therapy of beta-thalassemia), by Telethon (contract GGP10214) and AIRC (Italian Association for Cancer Research). This research is also supported by CIB (Interuniversity Consortium for Biotechnologies).

References

1. He L, Hannon GJ (2010) MicroRNAs: small RNAs with a big role in gene regulation. Nat Rev Genet 5:522–531
2. Manicardi A, Fabbri E, Tedeschi T, Sforza S, Bianchi N et al (2012) Cellular Uptakes, Biostabilities and Anti-miR-210 Activities of Chiral Arginine-PNAs in Leukaemic K562 Cells. Chembiochem 13:1327–1337
3. Fabbri E, Manicardi A, Tedeschi T, Sforza S, Bianchi N et al (2011) Modulation of the biological activity of microRNA-210 with peptide nucleic acids (PNAs). ChemMedChem 6: 2192–2202
4. Fabbri E, Brognara E, Borgatti M, Lampronti I, Finotti A et al (2011) miRNA therapeutics: delivery and biological activity of peptide nucleic acids targeting miRNAs. Epigenomics 3:733–745
5. Gambari R, Fabbri E, Borgatti M, Lampronti I, Finotti A et al (2011) Targeting microRNAs involved in human diseases: a novel approach for modification of gene expression and drug development. Biochem Pharmacol 82:1416–1429
6. Nielsen PE, Egholm M, Berg RH, Buchardt O (1991) Sequence-selective recognition of DNA by strand displacement with a thymine-substituted polyamide. Science 254:1497–1500
7. Nastruzzi C, Cortesi R, Esposito E, Gambari R, Borgatti M et al (2000) Liposomes as carriers for DNA-PNA hybrids. J Control Release 68:237–249
8. Karkare S, Bhatnagar D (2006) Promising nucleic acid analogs and mimics: characteristic features and applications of PNA, LNA, and morpholino. Appl Microbiol Biotechnol 71: 575–586
9. Tonelli R, Fronza S, Purgato S, Camerin C, Bologna F et al (2005) Anti-gene Peptide Nucleic Acid (PNA) Specifically Inhibits N-myc Expression in Human Neuroblastoma Cells Leading to Persistent Cell-growth Inhibition and Apoptosis. Mol Cancer Ther 4:779–786
10. Nielsen PE (2001) Targeting double stranded DNA with peptide nucleic acid (PNA). Curr Med Chem 8:545–550
11. Borgatti M, Lampronti I, Romanelli A, Pedone C, Saviano M et al (2003) Transcription factor decoy molecules based on a peptide nucleic acid (PNA)-DNA chimera mimicking Sp1 binding sites. J Biol Chem 278:7500–7509
12. Rasmussen FW, Bendifallah N, Zachar V, Shiraishi T, Fink T et al (2006) Evaluation of transfection protocols for unmodified and modified peptide nucleic acid (PNA) oligomers. Oligonucleotides 16:43–57
13. Cortesi R, Mischiati C, Borgatti M, Breda L, Romanelli A et al (2004) Formulations for

natural and peptide nucleic acids based on cationic polymeric submicron particles. AAPS Pharmsci 6:10–21

14. Borgatti M, Breda L, Cortesi R, Nastruzzi C, Romanelli A et al (2002) Cationic liposomes as delivery systems for double-stranded PNA-DNA chimeras exhibiting decoy activity against NF-kappaB transcription factors. Biochem Pharmacol 64:609–616
15. Abes R, Arzumanov A, Moulton H, Abes S, Ivanova G (2008) Arginine-rich cell penetrating peptides: design, structure-activity, and applications to alter pre-mRNA splicing by steric-block oligonucleotides. J Pept Sci 14:455–460
16. Torres AG, Threlfall RN, Gait MJ (2011) Potent and sustained cellular inhibition of miR-122 by lysine-derivatized peptide nucleic acids (PNA) and phosphorothioate locked nucleic acid (LNA)/2′-O-methyl (OMe) mixmer anti-miRs in the absence of transfection agents. Artificial DNA PNA XNA 3:71–78
17. Zhang C, Kang C, You Y, Pu P, Yang W (2009) Co-suppression of miR-221/222 cluster suppresses human glioma cell growth by targeting p27kip1 in vitro and in vivo. Int J Oncol 34: 1653–1660
18. Le Sage C, Nagel R, Egan DA, Schrier M, Mesman E (2007) Regulation of the p27(Kip1) tumor suppressor by miR-221 and miR-222 promotes cancer cell proliferation. EMBO J 26:3699–3708
19. Galardi S, Mercatelli N, Giorda E, Massalini S, Frajese GV, Ciafrè SA, Farace MG (2007) miR-221 and miR-222 expression affects the proliferation potential of human prostate carcinoma cell lines by targeting p27Kip1. J Biol Chem 282:2316–2324
20. Mizuma M, Katayose Y, Yamamoto K, Shiraso S, Sasaki T (2008) Up-regulated p27Kip1 reduces matrix metalloproteinase-9 and inhibits invasion of human breast cancer cells. Anticancer Res 28:2669–2677
21. Bianchi N, Zuccato C, Lampronti I, Borgatti M, Gambari R (2009) Expression of miR-210 during erythroid differentiation and induction of gamma-globin gene expression. BMB Rep 42:493–499
22. Brognara E, Fabbri E, Aimi F, Manicardi A, Bianchi N et al (2012) Peptide nucleic acids targeting miR-221 modulate p27^{Kip1} expression in breast cancer MDA-MB-231 cells. Int J Oncol 41:2119–2127.

Chapter 15

Lentiviral Overexpression of miRNAs

Hannah Zöllner, Stephan A. Hahn, and Abdelouahid Maghnouj

Abstract

Deregulation of microRNAs (miRNAs) has been attributed to almost any human disease analyzed to date. This calls for models and experimental strategies for functional analyses of miRNAs enabling miRNA overexpression or suppression in target cell and/or tissues. Lentiviral vector (LV)-based technologies allow the long-term overexpression of miRNAs in nearly all cell types in an easy and rapid manner and are therefore popular tools for this application. In this chapter we describe the cloning of LV miRNA expression vectors as well as the production of virus particles for target cell infection and stable expression of miRNAs.

Key words MicroRNA overexpression, Lentiviral vector, Cloning, Sh-miR, Pri-miR

1 Introduction

To study the function of specific miRNAs, overexpression or suppression techniques are commonly used to reach this goal. The currently most widely used miRNA mimic (miR-Mimic) strategy for miRNA overexpression is based on synthetic nonnatural double-stranded miRNA-like RNA. This approach is limited to transient expression of the miRNA under study. LV-based expression of miRNAs however allows the stable and long-term expression of an miRNA hairpin sequence in a large variety of cell types, including those that are difficult to transfect with standard technology such as primary or nondividing cells [1, 2].

MiRNA expression cassettes in LV can be designed as small hairpins (sh-miRs) or as primary miRNA (pri-miRs). Sh-miRs are perfect hairpins with a small apical loop and a 3′ UU-overhang. The 3′ end of the loop sequence contains the endogenous miR sequence (herein called sense sequence, *see* also Fig. 1) because it is preferred for producing the mature miRNA sequence, whereas the 5′ end of the loop sequence contains the fully complementary sequence to the mature miRNA (herein called anti-mature sequence) [3]. Sh-RNA synthesis systems are driven by RNA polymerase III (pol III) promoters such as variants of U6 or H1

Christoph Arenz (ed.), *miRNA Maturation: Methods and Protocols*, Methods in Molecular Biology, vol. 1095, DOI 10.1007/978-1-62703-703-7_15,

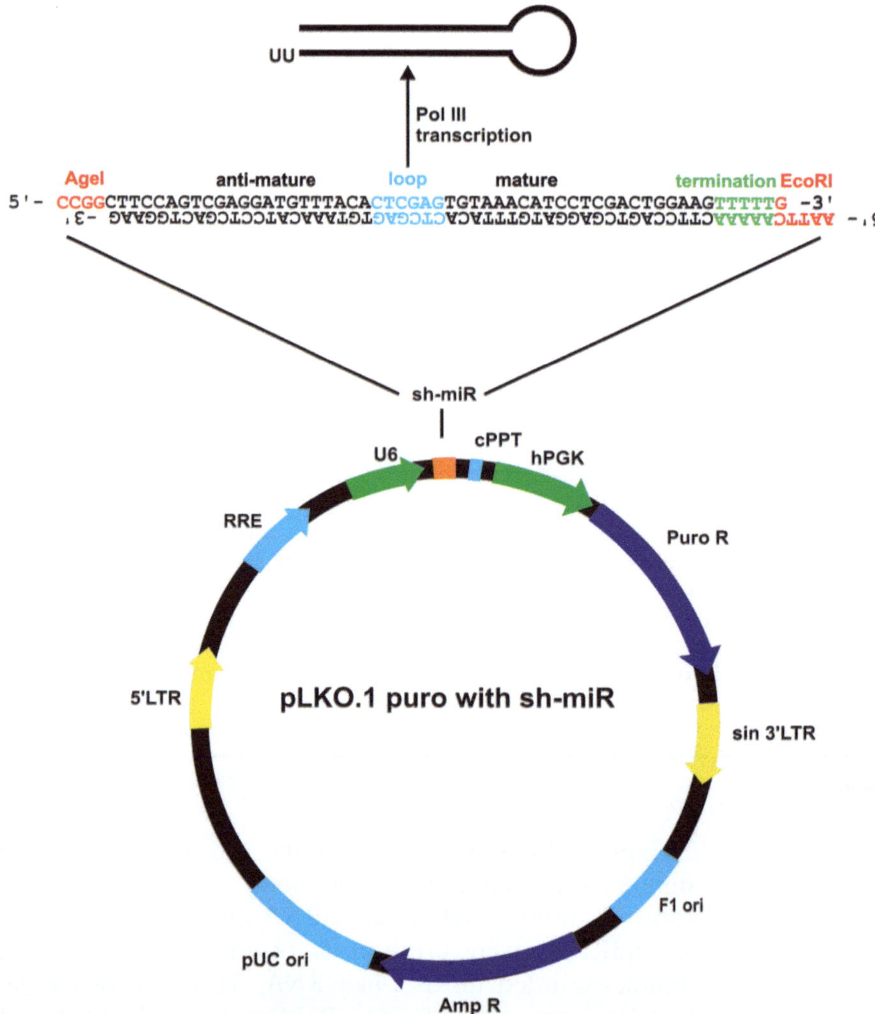

Fig. 1 Map of pLKO.1-puro vector containing a sh-miR insert (here hsa-miR-30a-5p). The U6 promoter directs RNA polymerase III transcription of the sh-miR. The sh-miR sequence is followed by a polyT termination signal for RNA polymerase III. *Mature* indicates the position of the mature miRNA sequence under study. *Anti-mature* indicates the sequence 100 % complementary to the miRNA under study

promoters [4]. Among the main advantages of the sh-miR approach is the ease of its cloning strategy which only requires the commercial synthesis of two complementary oligonucleotides. Expressing an miRNA as a pri-miR in turn requires the PCR amplification of a genomic fragment including the approximately 100 bp long precursor-miRNA (pre-miR) sequence flanked by some 200–300 bp upstream and downstream of the pre-miR sequence. Apart from decoupling the pri-miR expression from its endogenous promoter, the expression and maturation of vector-based pri-miRs include all endogenous processing steps and may thus better mimic natural miRNA biogenesis. Furthermore, both the endogenous so-called major and minor miRNAs are produced with this expression vector system. Thus, if the experimental goal requires the specific testing of only the minor or major miRNA sequence, the expression of an miRNA-sponge vector competing out the unwanted miRNA sequence may be entertained [3].

The LV system described herein is an HIV-based third generation three-vector system [1]. The first vector contains the LTRs (Long Terminal Repeats) required for reverse transcription and integration, the packaging signal ψ and the heterologous sequences (here miRNA sequence). The two remaining plasmids, the so-called helper plasmids, which provide all the structural proteins (HIV-1 Gag/Pol, Tat, Rev, and Env), are needed in order to package the new virus within the packaging cell [5]. Due to the ψ deletion within the helper plasmids, this transcribed RNA does not get incorporated into the new recombinant virus. Choosing the envelope (Env) protein for pseudotyping and thus controlling the tropism of the virus is important. The VSV-G protein (G glycoprotein of the vesicular stomatitis virus) is often used because it allows the transduction of a wide variety of cells [4]. For lentivirus production, the packaging cell line HEK 293T is transiently transfected with the three plasmids. Following transfer of the lentiviral particles generated by the HEK 293T cells to the target cells the VSV-G interact with a phospholipid component of the cell membrane. This induces the fusion of the membranes of the virus–envelope and the host cell. After the entry of the virus, uncoating and reverse transcription start immediately and subsequently the formation of a pre-integration complex (PIC) takes place. The PIC is a specific hallmark of lentiviruses, enabling them to actively enter the nucleus of a cell in order to infect nondividing cells. As soon as the viral DNA is integrated into the host cell genome, gene expression is initiated [6].

2 Material

The overexpression of miRNAs via lentiviral particles requires Biosafety Level 2 (BSL-2). Handle particles as a potentially biohazardous material and strictly follow all published BSL-2 guidelines. The transduced target cells can be handled as Biosafety Level 1 after several round of medium exchange following removal of the viral particles. Prepare all solutions using ultrapure water (18 MΩ cm at 25 °C) and sterile filtrate (0.2 μm pore size); aliquots are stored at −20 °C (unless indicated otherwise).

2.1 Suggested Vectors

1. For sh-miR expression: pLKO.1 puro (addgene no. 8453).
2. For pri-miR expression: pLJM1-EGFP (addgene no. 19319).
3. For HIV-1 GAG/POL, Tat and Rev expression (helper plasmid): pCMV delta R8.2 (addgene no. 12263).
4. For VSV-G expression (helper plasmid): pCMV-VSV-G (addgene no. 8454) or pHIT/G [7].

2.2 Cloning

1. TE buffer: 10 mM Tris–HCl (pH 8), 1 mM EDTA.
2. Enzymes:
   - restriction enzymes: *Eco*RI, *Age*I.
   - ligase, e.g., T4 DNA Ligase, 1 U/μL, Invitrogen.
   - proof-reading polymerase, e.g., Phusion® High-Fidelity DNA Polymerase, NEB.
   - Taq polymerase.
3. Gel extraction kit, e.g., peqGOLD Gel Extraction Kit (C-Line), Peqlab.
4. dNTPs (10 mM each) in water.
5. TAE buffer: 1 mM EDTA, 40 mM Tris (pH 8.3), 20 mM acetic acid. TAE buffer is commonly prepared as a 50× stock solution: dissolve 242 g Tris base in water, add 57.1 mL glacial acetic acid and 3.3 g of disodium ethylenediaminetetraacetate $\times 2H_2O$, and bring the final volume up to 1 l. This stock solution can be diluted 50:1 with water to make a 1× working solution.
6. SB (sodium boric acid) buffer: 5 mM disodium borate decahydrate. A 25× stock solution is prepared by dissolving 95.34 g $Na_2B_4O_7 \cdot 10\ H_2O$ in 2 L H_2O.
7. Agarose gels (1–1.5 %) in 1× TAE buffer: Weigh 1–1.5 g of agarose in a 250 mL Erlenmeyer flask (*see* **Note 1**) and boil it in 100 mL 1× TAE buffer. Add 1 μL ethidium bromide (10 mg/mL; Fluka) just before the solution is lukewarm, mix it cautiously and pour it into a gel casting tray (*see* **Note 2**). As running buffer use 1× TAE buffer.
8. Agarose gels (2 %) in 1× SB buffer: Boil 2 g of agarose in 100 mL SB buffer as in previous step.
9. 5× TAE loading buffer: Dissolve 45 g Ficoll 70, ~20 mg bromophenol blue and 2.25 g *N*-lauroylsarcosin in 50 mL H_2O. Add 3.6 mL glycerol and 9 mL 50× TAE buffer. Adjust the volume to 90 mL with H_2O.
10. 5× SB loading buffer: Dissolve 45 g Ficoll 70, ~20 mg xylene cyanol, and 2.25 g *N*-lauroylsarcosin in 50 mL H_2O. Add 3.6 mL glycerol and 18 mL 25× SB buffer. Adjust the volume to 90 mL with H_2O.
11. Antibiotics (bacteria): Dissolve ampicillin or carbenicillin in water at 100 mg/mL and store at −20 °C in aliquots.
12. LB-Medium: Suspend 20 g LB-Broth in 1 L of water and autoclave for 15 min at 121 °C. Storage at 4 °C or room temperature (*see* **Note 3**).
13. LB-agar: Suspend 705 g LB-Agar in 2 L water, aliquot 400 mL each with a stir bar in 0.5 L flasks and autoclave for

15 min at 121 °C. Prior to pour plates, melt the LB-Agar in a microwave (approx. 500 W) and add 1/1,000 volume ampicillin or carbenicillin (end concentration 100 μg/mL) just before the solution is lukewarm. Pour about 20 plates (storage at 4 °C in the dark).

14. Stbl3 E. coli strain is used for transformation (*see* **Note 4**). Stbl3 genotype: F–*mcr*B *mrr hsd*S20(rB–, mB–) *rec*A13 *sup*E44 *ara*-14 *gal*K2 *lac*Y1 *pro*A2 *rps*L20(StrR)*xyl*-5 λ– *leu mtl*-1.
15. PC8 (Phenol–Chloroform, pH 8).
16. 3 M Sodium acetate, pH 5.2: Dissolve 40.82 g sodium acetate-trihydrate in 80 mL water. Adjust to pH 5.2 with glacial acetic acid and fill volume up to 100 mL.
17. Glycogen 20 mg/mL (e.g., Roche).
18. Primers:
    - pLKO-forward:5′-GACTGTAAACACAAAGATATTAG-3′.
    - pLKO-reverse:5′-TAATTCTTTAGTTTGTATGTCTG-3′.
    - pLJM1-forward: 5′-TAACAACTCCGCCCCATTGA-3′.
19. 10× RDA buffer: 670 mM Tris (pH 8.8), 160 mM ammonium sulfate, 100 mM β-mercaptoethanol, 1 mg/mL BSA.
20. 50 mM $MgCl_2$: Dissolve 1.02 g $MgCl_2 \times 6H_2O$ in 100 mL.
21. Midi-Kit for purifying vectors after cloning, e.g., PureYield™ DNA Midiprep System (Promega).
22. Glycerol, 99.5 %.

2.3 Lentiviral Particles

1. 0.1 M $NaPO_4$ solution (pH 7.1): Dissolve 2.76 g NaH_2PO_4–$1H_2O$ in 100 mL H_2O. Dissolve 5.36 g Na_2HPO_4–$7H_2O$ in 100 mL H_2O. Mix both solutions (~1:2) to obtain pH 7.1 and double the volume with water.
2. 2× HBS buffer (HEPES buffered saline): 50 mM HEPES, 280 mM NaCl, 1.5 mM $NaHPO_4$. Dissolve 6.5 g HEPES, 8 g NaCl, and 7.5 mL 0.1 M $NaPO_4$ (pH 7.1) in 450 mL H_2O. Adjust pH exactly to 7.1, bring up to 500 mL with H_2O and aliquot in 50 mL tubes.
3. 2 M $CaCl_2$: 87.6 g $CaCl_2$-6×H_2O in a final volume of 200 mL water. Filter sterilize and store at 4 °C.
4. Polybrene (Hexadimthrine bromide): 4 mg/mL solution in water. The aliquot in use can be stored at 4 °C.
5. DMEM supplemented with 10 % fetal calf serum, 100 U/mL penicillin, and 100 μg/mL streptomycin.
6. Puromycin: Dissolve puromycin in water at 50 mg/mL and store at –20 °C in aliquots.
7. Phosphate-buffered saline (1× PBS, pH 7.4): commercial source.

8. Trypsin–EDTA: 0,05 %: commercial source.
9. HEK 293T cells (ATCC, Manassas, VA, USA).
10. Sterile filters (0.45 μm).
11. 20 mL syringe.

3 Methods

Due to the differences in the cloning procedures required for generation of pri-miR or sh-miR-expression vectors, experimental details are given separately for both strategies whenever necessary.

3.1 Design of Primers for pri-miRNAs

1. Pre-miRNA and flanking sequences can be obtained from http://useast.ensembl.org/index.html.
2. Design primers complementary to the ends of the pri-miR region (pre-microRNA (60–70 nucleotides) and their flanking 200–300 nucleotides) with the following features: Primer length between 18 and 24 nucleotides. The primers should be long enough to give a melting temperature (T_m) of 55–60 °C. T_ms for both primers should be within 2 °C. The GC-content should be 40–60 %. The 8 bases at the 3′ end should preferably have 5 or more As and Ts to prevent it from being sticky. The 3′ end itself should be a G or C to improve the efficiency of priming. The second and third bases from the 3′ end should be As and/or Ts.
3. Primer pairs should be checked for complementarity at the 3′-end to reduce primer–dimer formation (no more than 3 to max. 4 nt complementary to any primer sequence).
4. Extend the 5′-end (forward) primer with the *Age*I and the 3′-end (reverse) primer with the *Eco*RI recognition site.
5. To increase cleavage efficiency, add 4 bases to the 5′-end of each restriction site. This is necessary because restriction enzymes cleave DNA much less efficient towards the end of a fragment. An example is shown below:
   - Forward primer with *Age*I: 5′-CGCTACCGGTNNNN NNNNNNNNNNNNNNNN-3′.
   - Reverse primer with *Eco*RI: 5′-TCGAGAATTCNNNNNN NNNNNNNNNNNNNN-3′.

3.2 Design of Primers for sh-RNAs

1. Mature miRNA sequence can be obtained from http://www.mirbase.org/
2. Design and synthesize two complementary single-stranded oligonucleotides as illustrated in Fig. 1 (*see* **Note 5**): The sense strand oligo contains an *Age*I overhang (CCGG) at the 5′ end followed by the anti-mature-, loop-, and mature sequence and

is terminated with TTTTT sequence which constitutes the pol III terminator (Fig. 1). The antisense strand oligo is with the exception of its *Eco*RI overhang (AATT) at the 5′ end 100 % complementary to the sense oligo.

4 Generation of Inserts by Amplification of Genomic Regions Containing a pri-miR Sequence

1. Amplify the pri-miR from (human) genomic DNA using primers with *Age*I and *Eco*RI adaptors. Use a proof-reading DNA polymerase.
2. Review outcome (specificity) of the PCR by applying an aliquot to an agarose gel (1.5 %).
3. Purify PCR reaction from polymerase using standard phenol–chloroform extraction and exchange buffer by standard precipitation (*see* Subheading 4.2).
4. Digest the PCR product with *Age*I and *Eco*RI.
5. Gel-purify the amplicon of the pri-miR.

4.1 Generation of Inserts by Hybridization of Oligonucleotides

1. Dissolve the lyophilized molecules in TE to a final concentration of 100 μM.
2. Mix complementary oligonucleotides together at a 1:1 M ratio in a 0.5 mL reaction tube:
   - 1.5 μL sense oligo.
   - 1.5 μL antisense oligo.
   - 10 μL 5× ligase buffer.
   - 37 μL H_2O (*see* **Note 6**).
3. Perform the annealing reaction in a PCR-cycler using the following protocol (*see* **Note 7**):

99 °C	1 min
96 °C	7 min
85 °C	5 min
75 °C	7 min
65 °C	10 min
37 °C	10 min
22 °C	20 min
16 °C	10 min
6 °C	Hold

4. Review each probe on a 2 % SB-gel. This allows estimation of the concentration of hybridized oligonucleotides (Fig. 2).

4.2 Ligation

1. Digest 5–10 μg plasmid DNA (pLKO1.puro/pLJM1) with *Age*I and *Eco*RI.
2. Purify digested plasmid DNA using a gel extraction kit and determine its concentration.
3. Combine 50–100 ng of vector with a three- to tenfold molar excess (*see* **Note 8**) of insert. The amount of insert can be calculated using the following formula:

$$ng\ insert = \left[\frac{size\ insert\ in\ bp \times ng\ vector}{size\ vector\ in\ bp}\right] \times ratio$$

4. Plan control reactions: One reaction without insert and one without enzyme.

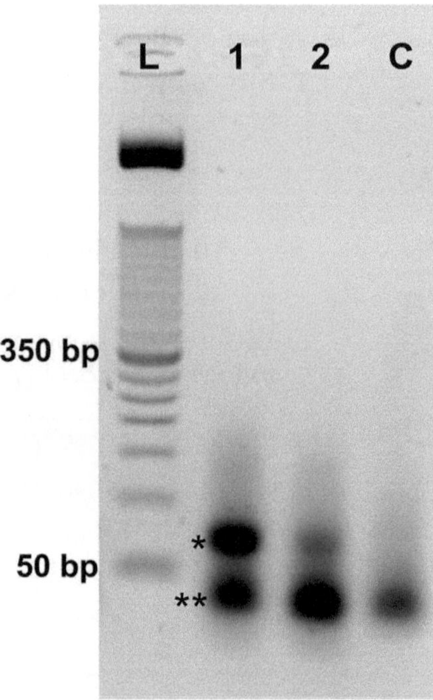

Fig. 2 Hybridization efficiency of oligonucleotides was analyzed by 2 % SB agarose gel electrophoresis. The proportion of single versus double-stranded oligonucleotides can be roughly estimated from the staining intensity or the corresponding bands. *Asterisk* indicates double-stranded oligonucleotide; *double asterisk* indicates single-stranded oligonucleotides; *L* 50-bp DNA ladder (Invitrogen), *C* negative control (only sense oligonucleotide), *1* and *2* result of two representative oligonucleotide hybridization experiments (*see* **Note 23**)

5. We recommend using a total volume of 20 μL for the ligase reaction. Use 1 U ligase for each ligation. Mix all components by pipetting.
6. Incubate for 1 h at room temperature or at 16 °C overnight (*see* **Note 9**).
7. Remove the proteins from the solution by phenol–chloroform extraction: Bring the ligation mixture up to 200 μL with H_2O, then add an equal volume of PC8 to the DNA containing reaction mixture and vortex (*see* **Note 10**). Then separate the aqueous phase which contains the DNA from the organic phase by centrifugation at 16,000 × *g* for 5 min. Lastly, transfer the aqueous phase with care into a fresh (labeled) reaction tube.
8. Ethanol-precipitate the nucleic acid by adding 0.1 volume of 3 M sodium acetate, pH 5.2 and 2 μL of glycogen to the aqueous phase, vortex, then add 2 volumes of absolute ethanol and vortex again. Recover the precipitated DNA by centrifugation at 16,000 × *g* for 5 min. Then, remove the ethanol with care (*see* **Note 11**) and wash the pellet three times with 70 % ethanol. Finally, dry the pellet at room temperature or 37 °C and dissolve it in 10 μL H_2O.

5 Transformation of Electroporation-Competent *E. coli*

1. Thaw electroporation-competent bacteria cells on ice (we recommend the use of bacteria with a transformation efficiency of at least 1×10^8 cfu/μg).
2. Chill 2–3 μL of the ligation mixture in a 1.5 mL reaction tube.
3. Add competent cells to the DNA. Mix gently by pipetting or flicking the tube 4–5 times to mix the cells and DNA. Do not vortex.
4. Transfer the DNA–cell mixture into a chilled electroporation cuvette (0.1 cm gap).
5. Put the dry cuvette into the sample chamber of an electroporator and pulse once at 1,800 V.
6. Remove the cuvette from the sample chamber and add 1 mL of LB-medium immediately (*see* **Note 12**).
7. Transfer the cell suspension into a 2 mL reaction tube.
8. Incubate the cell suspension for 45–60 min at 37 °C while shaking at 190–220 rpm.
9. Plate aliquots of the bacteria suspension (*see* **Note 13**) on LB-agar plates supplemented with antibiotics.
10. Incubate plates at 37 °C overnight.

5.1 Direct Bacteria PCR

Direct bacteria PCR is a technique which allows selecting clones from transformed bacteria cells that contain the desired DNA without prior plasmid preparation and enzymatic tests.

1. Pick at least 10 individual colonies. Pick also 2 colonies form the control plate.
2. Grow colonies for at least 4 h (37 °C, 190–220 rpm) in 400 μL LB-medium containing 100 μg/mL ampicillin/carbenicillin in a 1.5 mL reaction tube.
3. For PCR mix the following components on ice together; always adding enzyme last. For multiple samples, make a master mix and aliquot 10 μL in each well of a PCR-plate (also on ice):

10× RDA buffer	1.5 μL
$MgCl_2$	1.2 μL
dNTPs (10 mM each)	0.3 μL
Primer sense (10 μM) (pLKO-forward/pLJM1-forward)	0.7 μL
Primer antisense (10 μM) (pLKO-reverse)	0.7 μL
H_2O	4.6 μL
Taq	1 μL

4. Add 5 μL of the bacteria suspension. Don't forget controls (H_2O instead of bacteria suspension).
5. Cover wells with mineral oil.
6. Use the following PCR conditions:

95 °C	3 min	
95 °C	30 s	30x
53 °C	30 s	30x
72 °C	30–45 s	30x
72 °C	5 min	

7. To check the outcome, apply 5 μL of each PCR reaction to a 1.5 % agarose gel.

5.2 Preparation of Plasmid DNA

1. Set up overnight cultures of positive candidate colonies.
2. Purify the miRNA construct using an endotoxin-free plasmid purification kit (*see* **Note 14**).
3. Confirm the insert by sequence analysis of the miRNA construct using the direct bacteria PCR primers (*see* **Note 15**).

5.3 Generation of Lentiviral Particles

This protocol is designed for virus production in 10 cm culture plates or T-75 cm^2 flasks.

1. (Day 1) Trypsinize HEK 293T (packaging cells), count these cells and add 3×10^6 HEK 293T cells to a 10 cm culture plate in 10 mL of pre-warmed medium (*see* **Note 16**).
2. Cell density should be approx. 50 % confluent for transfection (*see* **Note 17**).
3. (Day 2) Remove 2× HBS buffer and vectors from −20 °C freezer and pre-warm at room temperature.
4. Prepare 1 mL of calcium phosphate–DNA suspension for each 10 cm plate of cells as follows: Mix 12 μg of the transfer vector pLKO1.puro or pLJM1 containing the transgene (*see* **Note 18**), 12 μg of pCMVΔR8.2 and 6 μg of pCMV-VSV-G or pHit/G and fill it up with water to a total volume of 438 μL. Then add 62 μL of 2 M CaCl2 and mix by pipetting. Next add 500 μL of 2× HBS dropwise while holding the solution motionless, then incubate the mixture at room temperature for 10 min.
5. Mix 12 μg of the transfer vector pLKO1.puro or pLJM1 containing the transgene (*see* **Note 18**), 12 μg of pCMVΔR8.2 and 6 μg of pCMV-VSV-G or pHit/G and fill it up with water to a total volume of 438 μL. Then add 62 μL of 2 M $CaCl_2$ and mix by pipetting. Add 500 μL of 2× HBS dropwise while holding the solution motionless and incubate the mixture at room temperature for 10 min.
6. Add the 1 mL of calcium phosphate–DNA suspension dropwise into the medium of the HEK 293T cells (*see* **Notes 19** and **20**).
7. Incubate the cells at 37 °C, 10 % CO_2.
8. (Day 3) After 14–16 h, discard the culture medium from the HEK 293T cells and replace with 11 mL pre-warmed fresh medium to produce virions overnight. The transfection efficiency can be assessed by fluorescence microscopy of packaging cells transfected with the GFP-expressing control vector.
9. Plate an appropriate number of target cells, so that confluence reaches 50–60 % the next day. For 10 cm plates the cell number should be about 1×10^6, for 6 wells 1×10^5 cells are sufficient.
10. (Day 4) Remove the 11 mL virus-containing medium from the HEK 293T cells by a 20 mL syringe and filter into a 50 mL tube through a sterile filter (pore size 0.45 μm) to remove cell debris. To enhance infection, add polybrene in a final concentration of 4 μg/mL (*see* **Note 21**).
11. Dispose the packaging cells.

12. Aspirate the medium from the target cells and subsequently add the virus-containing medium. If target cells were seeded on a 10 cm plate use 10 mL, for 6 wells use at least 2 mL.
13. Incubate the target cells at 37 °C, 10 % CO_2 overnight (S2).
14. (Day 5) Remove the virus-containing medium from the target cells, wash twice with PBS and supply fresh medium.
15. If a vector expressing a living color reporter gene such as EGFP was used, expression of the reporter can be monitored 24 h after infection via fluorescence microscopy. A EGFP lentiviral vector provides not only a useful control for transient transfection efficiency of packaging HEK 293T cells but also for the target cell transduction efficiency.
16. If required, start selecting the cells using the minimum concentration of puromycin that causes complete cell death within 5–10 days (*see* **Note 22**).
17. MiRNA expression can be monitored, i.e., via qRT-PCR as early as 6 h post transduction and reaches maximum expression levels after 24 h.

6 Notes

1. It is advisable to use a large container, as long as it fits in the microwave to reduce the chance of boiling over agarose. Be aware of boiling delay.
2. Pour the gel slowly because in this case most bubbles stay up in the flask. If any bubbles are formed, push them to the side using a disposable tip.
3. We recommend storing it at room temperature. It is best flaming the bottle each time it is opened.
4. Use Stbl3™ *E. coli* for transformation, as this strain is particularly well-suited for cloning unstable DNA such as lentiviral DNA containing direct repeats.
5. MiRNA sequences with more than three Ts or As are not compatible with the sh-miR expression strategy, because premature termination of the transcript will be induced by the Ts.
6. In this mix the final concentration for the oligonucleotides reaches approx. 107 ng/μL.
7. Annealing occurs most efficiently when the temperature is slowly decreased after denaturation because the single oligonucleotides form a hairpin structure.
8. For ligation optimal results, use a 10:1 M ratio of sh-miR insert:vector and a 3–5:1 M ratio of pri-miR amplicon:vector.
9. Ligation at 16 °C overnight may result in a higher yield of colonies.

10. Phenol is a hazardous organic solvent. Always use suitable laboratory gloves when handling phenol containing solutions. Specific waste procedures may be required for the disposal of phenol-containing solutions.
11. Carefully pour out or aspirate supernatant (watch the DNA-pellet, do not lose it!).
12. Failure to immediately add medium to the electroporated cells can significantly reduce cell viability and decrease transformation efficiency.
13. It is useful to plate at least two different aliquots of bacterial suspension (i.e., 50 and 150 μL), in order to increase the chance to obtain enough single colonies the next day.
14. Don't forget to prepare a glycerol stock for maintaining the bacteria containing the desired plasmid.
15. The flanking sequence of a pri-miRNA clone may differ from the NCBI reference with respect to biological polymorphisms. This should not affect the function of the mature miRNA.
16. It is very important to have good single cell suspensions (trypsinize well) and to evenly distribute the cells.
17. Prior to use for lentiviral vector production, cells should have undergone at least 2 passages following thawing.
18. If the chosen vector does not contain a reporter gene such as EGFP for an easy qualitative estimate of the transfection and transduction efficiency of your viral preparations we highly recommend running a parallel experiment with a lentiviral vector of your choice expressing such a reporter. This will enable you to control all major steps of the protocol. It cannot control for transfection efficiency of your plasmid containing the transgene.
19. HEK 293T cells detach easily, be careful, as well with all medium changes.
20. From this step on, cells are considered as Biosafety Level 2.
21. We recommend not storing viral stock at this step, i.e., at −80 °C, because viral titers will significantly go down after one freeze thaw cycle. Furthermore, consider one of the many lentiviral titration methods in order to be able to compare virus production efficiency from experiment to experiment [8].
22. Prior to beginning experiments, determine the concentration of puromycin for target cells by performing a puromycin kill curve. Excess of puromycin can cause many undesired phenotypic responses in most cell types.
23. The agarose gel is non-denaturing; therefore, the single-stranded oligos do not resolve at the expected size due to formation of secondary structure.

References

1. Dull T, Zufferey R, Kelly M, Mandel RJ, Nguyen M, Trono D, Naldini L (1998) A third-generation lentivirus vector with a conditional packaging system. J Virol 72: 8463–8471
2. Naldini L (1998) Lentiviruses as gene transfer agents for delivery to non-dividing cells. Curr Opin Biotechnol 9:457–463
3. Baraniskin A, Birkenkamp-Demtroder K, Maghnouj A, Zollner H, Munding J, Klein-Scory S, Reinacher-Schick A, Schwarte-Waldhoff I, Schmiegel W, Hahn SA (2012) MiR-30a-5p suppresses tumor growth in colon carcinoma by targeting DTL. Carcinogenesis 33:732–739
4. Liu YP, Berkhout B (2011) miRNA cassettes in viral vectors: problems and solutions. Biochim Biophys Acta 1809:732–745
5. Klages N, Zufferey R, Trono D (2000) A stable system for the high-titer production of multiply attenuated lentiviral vectors. Mol Ther 2: 170–176
6. Engelman A, Cherepanov P (2012) The structural biology of HIV-1: mechanistic and therapeutic insights. Nat Rev Microbiol 10: 279–290
7. Fouchier RA, Meyer BE, Simon JH, Fischer U, Malim MH (1997) HIV-1 infection of non-dividing cells: evidence that the amino-terminal basic region of the viral matrix protein is important for Gag processing but not for post-entry nuclear import. EMBO J 16:4531–4539
8. Geraerts M, Willems S, Baekelandt V, Debyser Z, Gijsbers R (2006) Comparison of lentiviral vector titration methods. BMC Biotechnol 6:34

Index

Christoph Arenz (ed.), *miRNA Maturation: Methods and Protocols*, Methods in Molecular Biology, vol. 1095, DOI 10.1007/978-1-62703-703-7, © Springer Science+Business Media New York 2014

O

P

Q

R

S

T

W

Z

MIX
Papier aus verantwortungsvollen Quellen
Paper from responsible sources
FSC® C105338

If you have any concerns about our products,
you can contact us on
ProductSafety@springernature.com

In case Publisher is established outside the EU,
the EU authorized representative is:
Springer Nature Customer Service Center GmbH
Europaplatz 3, 69115 Heidelberg, Germany

Printed by Libri Plureos GmbH
in Hamburg, Germany